The Cumberbatch Tele-Accessibility Index related to CA, CT, FL, DC, IL, IN, MA, MI, NJ, NY, TN, and various other U.S. States - Years 2000-2007!

The information was obtained from The U.S. Census Bureau, and the Federal Communications Commission (FCC) of The United States of America.

Table of Contents

Executive Summary

The conclusion of this research paper is that there does exist a positive correlation between the level of tele-accessibility and the rate of income per capita, which was experienced between the states of **California**, **Connecticut**, **District of Columbia**, **Florida, Illinois, Indiana, Massachusetts, Michigan, New Jersey, New York, Pennsylvania, Tennessee, Virginia, Utah,** and **Ohio**, for the time period between 2000-2007.

Definitions

To obtain a better understanding of the paper I will provide definitions of various words which will be used throughout the document at the beginning, and are stated below.

The term **Fixed Lines** is a phrase which is used to describe the number of land-line or telephone lines that an individual has in their home, office, place of business, or in various other locations.

The term **Mobile Lines** is a phrase which is used to describe the number of mobile or cellular (cell) phones that an individual has, which may be used for personal, business or for numerous other purposes.

The term **Tele-Accessibility** is a phrase which is used to describe the total number of fixed plus mobile lines for a given year, which is then divided by the number of the population for that same time period. For example if a state has a tele-accessibility level of 1.7618, this means that the average household has access to 1.7618 means of communicating, i.e. a fixed and mobile telephone line, a fixed line, a mobile line, or some variation of these two terms.

The term **Income per Capita/10,000** is a phrase which I created so that one could more easily compare the average income per capita with the level of tele-accessibility experienced at that point in time. For the tele-accessibility numbers are relatively small and therefore to compare this variable with the income per capita without making their sizes relatively similar in magnitude would be challenging, and thus may minimize or diminish the over-all effect of the paper.

The term **Correlation Coefficient** is a phrase which is used to describe the extent to which two independent variables i.e. income per capita and tele-accessibility move in a similar direction when compared with each other. Therefore if they are positively correlated then one would expect that as the level of tele-accessibility increases or rises in a specific year, one would also be able to observe the level of income per capita also becoming greater or larger, for that time period as well.

The level of Tele-Accessibility in the year 2000

The state of **District of Columbia** had the highest level of tele-accessibility for the year

2000, at a rate of 2.31087, a measurement which was rounded off to five decimal places.

This means that on average 231.087% of the residents of **District of Columbia** at that

time period had access to various forms of telecommunication devices, i.e. fixed and

mobile phone line services. The level of income per capita for that same time period was

equal to **$37,383.00**. The correlation coefficient measurement for these two variables is

equal to 0.47636 (rounded to 5 decimal places), this means that 47.636% of the reason

for which one variable moves in a specific direction can be explained by the variation in

the other variable i.e. based upon the statistics provided 47.636% of the variation in the

level of income per capita in the state of **District of Columbia** at that time period could

be explained by the variation in the level of tele-accessibility, or access to

telecommunications.

The state of **New Jersey** had the second highest level of tele-accessibility for the year

2000, at a rate of 1.15648, a measurement which was rounded off to five decimal places.

This means that on average 115.648% of the residents of **New Jersey** at that time period

had access to various forms of telecommunication devices, i.e. fixed and mobile phone

line services. The level of income per capita for that same time period was equal to

$36,983.00. The correlation coefficient measurement for these two variables is equal to

0.47636 (rounded to 5 decimal places), this means that 47.636% of the reason for which one variable moves in a specific direction can be explained by the variation in the other variable i.e. based upon the statistics provided 47.636% of the variation in the level of income per capita in the state of **New Jersey** at that time period could be explained by the variation in the level of tele-accessibility, or access to telecommunications

The state of **California** had the third highest level of tele-accessibility for the year 2000, at a rate of 1.08951, a measurement which was rounded off to five decimal places. This means that on average 108.951% of the residents of **California** at that time period had access to various forms of telecommunication devices, i.e. fixed and mobile phone line services. The level of income per capita for that same time period was equal to **$32,275.00**. The correlation coefficient measurement for these two variables is equal to 0.47636 (rounded to 5 decimal places), this means that 47.636% of the reason for which one variable moves in a specific direction can be explained by the variation in the other variable i.e. based upon the statistics provided 47.636% of the variation in the level of income per capita in the state of **California** at that time period could be explained by the variation in the level of tele-accessibility, or access to telecommunications

The state of **Massachusetts** had the fourth highest level of tele-accessibility for the year 2000, at a rate of 1.08859, a measurement which was rounded off to five decimal places. This means that on average 108.859% of the residents of **Massachusetts** at that time period had access to various forms of telecommunication devices, i.e. fixed and mobile phone line services. The level of income per capita for that same time period was equal to **$37,992.00**. The correlation coefficient measurement for these two variables is equal to 0.47636 (rounded to 5 decimal places), this means that 47.636% of the reason for which one variable moves in a specific direction can be explained by the variation in the other variable i.e. based upon the statistics provided 47.636% of the variation in the level of income per capita in the state of **Massachusetts** at that time period could be explained by the variation in the level of tele-accessibility, or access to telecommunications.

The state of **Connecticut** had the fifth highest level of tele-accessibility for the year 2000, at a rate of 1.08767, a measurement which was rounded off to five decimal places. This means that on average 108.767% of the residents of **Connecticut** at that time period had access to various forms of telecommunication devices, i.e. fixed and mobile phone line services. The level of income per capita for that same time period was equal to **$40,640.00**. The correlation coefficient measurement for these two variables is equal to 0.47636 (rounded to 5 decimal places), this means that 47.636% of the reason for which one variable moves in a specific direction can be explained by the variation in the other

variable i.e. based upon the statistics provided 47.636% of the variation in the level of income per capita in the state of **Connecticut** at that time period could be explained by the variation in the level of tele-accessibility, or access to telecommunications.

The state of **Florida** had the sixth highest level of tele-accessibility for the year 2000, at a rate of 1.06486, a measurement which was rounded off to five decimal places. This means that on average 106.486% of the residents of **Florida** at that time period had access to various forms of telecommunication devices, i.e. fixed and mobile phone line services. The level of income per capita for that same time period was equal to **$28,145.00**. The correlation coefficient measurement for these two variables is equal to 0.47636 (rounded to 5 decimal places), this means that 47.636% of the reason for which one variable moves in a specific direction can be explained by the variation in the other variable i.e. based upon the statistics provided 47.636% of the variation in the level of income per capita in the state of **Florida** at that time period could be explained by the variation in the level of tele-accessibility, or access to telecommunications.

The state of **Illinois** had the seventh highest level of tele-accessibility for the year 2000, at a rate of 1.04921, a measurement which was rounded off to five decimal places. This means that on average 104.921% of the residents of **Illinois** at that time period had access to various forms of telecommunication devices, i.e. fixed and mobile phone line services. The level of income per capita for that same time period was equal to **$32,259.00**. The correlation coefficient measurement for these two variables is equal to 0.47636 (rounded to 5 decimal places), this means that 47.636% of the reason for which one variable moves in a specific direction can be explained by the variation in the other variable i.e. based upon the statistics provided 47.636% of the variation in the level of income per capita in the state of **Illinois** at that time period could be explained by the variation in the level of tele-accessibility, or access to telecommunications.

The state of **Pennsylvania** had the eighth highest level of tele-accessibility for the year 2000, at a rate of 1.03554, a measurement which was rounded off to five decimal places. This means that on average 103.554% of the residents of **Pennsylvania** at that time period had access to various forms of telecommunication devices, i.e. fixed and mobile phone line services. The level of income per capita for that same time period was equal to **$29,539.00**. The correlation coefficient measurement for these two variables is equal to 0.47636 (rounded to 5 decimal places), this means that 47.636% of the reason for which one variable moves in a specific direction can be explained by the variation in the other variable i.e. based upon the statistics provided 47.636% of the variation in the level of income per capita in the state of **Pennsylvania** at that time period could be explained by the variation in the level of tele-accessibility, or access to telecommunications.

The state of **Michigan** had the ninth highest level of tele-accessibility for the year 2000, at a rate of 1.01913, a measurement which was rounded off to five decimal places. This means that on average 101.913% of the residents of **Michigan** at that time period had access to various forms of telecommunication devices, i.e. fixed and mobile phone line services. The level of income per capita for that same time period was equal to **$29,612.00**. The correlation coefficient measurement for these two variables is equal to 0.47636 (rounded to 5 decimal places), this means that 47.636% of the reason for which one variable moves in a specific direction can be explained by the variation in the other variable i.e. based upon the statistics provided 47.636% of the variation in the level of income per capita in the state of **Michigan** at that time period could be explained by the variation in the level of tele-accessibility, or access to telecommunications.

The state of **New York** had the tenth highest level of tele-accessibility for the year 2000, at a rate of 0.98465, a measurement which was rounded off to five decimal places. This means that on average 98.465% of the residents of **New York** at that time period had access to various forms of telecommunication devices, i.e. fixed and mobile phone line services. The level of income per capita for that same time period was equal to **$34,547.00**. The correlation coefficient measurement for these two variables is equal to 0.47636 (rounded to 5 decimal places), this means that 47.636% of the reason for which one variable moves in a specific direction can be explained by the variation in the other variable i.e. based upon the statistics provided 47.636% of the variation in the level of

income per capita in the state of **New York** at that time period could be explained by the variation in the level of tele-accessibility, or access to telecommunications.

The state of **Virginia** had the eleventh highest level of tele-accessibility for the year 2000, at a rate of 0.97368, a measurement which was rounded off to five decimal places. This means that on average 97.368% of the residents of **Virginia** at that time period had access to various forms of telecommunication devices, i.e. fixed and mobile phone line services. The level of income per capita for that same time period was equal to **$31,162.00**. The correlation coefficient measurement for these two variables is equal to 0.47636 (rounded to 5 decimal places), this means that 47.636% of the reason for which one variable moves in a specific direction can be explained by the variation in the other variable i.e. based upon the statistics provided 47.636% of the variation in the level of income per capita in the state of **Virginia** at that time period could be explained by the variation in the level of tele-accessibility, or access to telecommunications.

The state of **Tennessee** had the twelfth highest level of tele-accessibility for the year 2000, at a rate of 0.94716, a measurement which was rounded off to five decimal places. This means that on average 94.716% of the residents of **Tennessee** at that time period had access to various forms of telecommunication devices, i.e. fixed and mobile phone line services. The level of income per capita for that same time period was equal to

$26,239.00. The correlation coefficient measurement for these two variables is equal to 0.47636 (rounded to 5 decimal places), this means that 47.636% of the reason for which one variable moves in a specific direction can be explained by the variation in the other variable i.e. based upon the statistics provided 47.636% of the variation in the level of income per capita in the state of **Tennessee** at that time period could be explained by the variation in the level of tele-accessibility, or access to telecommunications.

The state of **Ohio** had the thirteenth highest level of tele-accessibility for the year 2000, at a rate of 0.92310, a measurement which was rounded off to five decimal places. This means that on average 92.310% of the residents of **Ohio** at that time period had access to various forms of telecommunication devices, i.e. fixed and mobile phone line services. The level of income per capita for that same time period was equal to **$28,400.00**. The correlation coefficient measurement for these two variables is equal to 0.47636 (rounded to 5 decimal places), this means that 47.636% of the reason for which one variable moves in a specific direction can be explained by the variation in the other variable i.e. based upon the statistics provided 47.636% of the variation in the level of income per capita in the state of **Ohio** at that time period could be explained by the variation in the level of tele-accessibility, or access to telecommunications.

The state of **Indiana** had the fourteenth highest level of tele-accessibility for the year 2000, at a rate of 0.89812, a measurement which was rounded off to five decimal places. This means that on average 89.812% of the residents of **Indiana** at that time period had access to various forms of telecommunication devices, i.e. fixed and mobile phone line services. The level of income per capita for that same time period was equal to **$27,011.00**. The correlation coefficient measurement for these two variables is equal to 0.47636 (rounded to 5 decimal places), this means that 47.636% of the reason for which one variable moves in a specific direction can be explained by the variation in the other variable i.e. based upon the statistics provided 47.636% of the variation in the level of income per capita in the state of **Indiana** at that time period could be explained by the variation in the level of tele-accessibility, or access to telecommunications.

The state of **Utah** had the lowest level of tele-accessibility for the year 2000, at a rate of 0.88162, a measurement which was rounded off to five decimal places. This means that on average 88.162% of the residents of **Utah** at that time period had access to various forms of telecommunication devices, i.e. fixed and mobile phone line services. The level of income per capita for that same time period was equal to **$23,907.00**. The correlation coefficient measurement for these two variables is equal to 0.47636 (rounded to 5 decimal places), this means that 47.636% of the reason for which one variable moves in a specific direction can be explained by the variation in the other variable i.e. based upon the statistics provided 47.636% of the variation in the level of income per capita in the

state of **Utah** at that time period could be explained by the variation in the level of tele-

accessibility, or access to telecommunications.

The level of Tele-Accessibility in the year 2001

The state of **District of Columbia** had the highest level of tele-accessibility for the year 2001, at a rate of 2.25251, a measurement which was rounded off to five decimal places. This means that on average 225.251% of the residents of **District of Columbia** at that time period had access to various forms of telecommunication devices, i.e. fixed and mobile phone line services. The level of income per capita for that same time period was equal to **$40,150.00**. The correlation coefficient measurement for these two variables is equal to 0.56317 (rounded to 5 decimal places), this means that 56.317% of the reason for which one variable moves in a specific direction can be explained by the variation in the other variable i.e. based upon the statistics provided 56.317% of the variation in the level of income per capita in the state of **District of Columbia** at that time period could be explained by the variation in the level of tele-accessibility, or access to telecommunications.

The state of **New Jersey** had the second highest level of tele-accessibility for the year 2001, at a rate of 1.27454, a measurement which was rounded off to five decimal places. This means that on average 127.454% of the residents of **New Jersey** at that time period had access to various forms of telecommunication devices, i.e. fixed and mobile phone line services. The level of income per capita for that same time period was equal to

$38,509.00. The correlation coefficient measurement for these two variables is equal to 0.56317 (rounded to 5 decimal places), this means that 56.317% of the reason for which one variable moves in a specific direction can be explained by the variation in the other variable i.e. based upon the statistics provided 56.317% of the variation in the level of income per capita in the state of **New Jersey** at that time period could be explained by the variation in the level of tele-accessibility, or access to telecommunications

The state of **Florida** had the third highest level of tele-accessibility for the year 2001, at a rate of 1.15291, a measurement which was rounded off to five decimal places. This means that on average 115.291% of the residents of **Florida** at that time period had access to various forms of telecommunication devices, i.e. fixed and mobile phone line services. The level of income per capita for that same time period was equal to **$28,947.00**. The correlation coefficient measurement for these two variables is equal to 0.56317 (rounded to 5 decimal places), this means that 56.317% of the reason for which one variable moves in a specific direction can be explained by the variation in the other variable i.e. based upon the statistics provided 56.317% of the variation in the level of income per capita in the state of **Florida** at that time period could be explained by the variation in the level of tele-accessibility, or access to telecommunications

The state of **Massachusetts** had the fourth highest level of tele-accessibility for the year 2001, at a rate of 1.11734, a measurement which was rounded off to five decimal places. This means that on average 111.734% of the residents of **Massachusetts** at that time period had access to various forms of telecommunication devices, i.e. fixed and mobile phone line services. The level of income per capita for that same time period was equal to **$ 38,907.00**. The correlation coefficient measurement for these two variables is equal to 0.56317 (rounded to 5 decimal places), this means that 56.317% of the reason for which one variable moves in a specific direction can be explained by the variation in the other variable i.e. based upon the statistics provided 56.317% of the variation in the level of income per capita in the state of **Massachusetts** at that time period could be explained by the variation in the level of tele-accessibility, or access to telecommunications.

The state of **Connecticut** had the fifth highest level of tele-accessibility for the year 2001, at a rate of 1.11569, a measurement which was rounded off to five decimal places. This means that on average 111.569% of the residents of **Connecticut** at that time period had access to various forms of telecommunication devices, i.e. fixed and mobile phone line services. The level of income per capita for that same time period was equal to **$42,435.00**. The correlation coefficient measurement for these two variables is equal to 0.56317 (rounded to 5 decimal places), this means that 56.317% of the reason for which one variable moves in a specific direction can be explained by the variation in the other

variable i.e. based upon the statistics provided 56.317% of the variation in the level of income per capita in the state of **Connecticut** at that time period could be explained by the variation in the level of tele-accessibility, or access to telecommunications.

The state of **Illinois** had the sixth highest level of tele-accessibility for the year 2001, at a rate of 1.09003, a measurement which was rounded off to five decimal places. This means that on average 109.003% of the residents of **Illinois** at that time period had access to various forms of telecommunication devices, i.e. fixed and mobile phone line services. The level of income per capita for that same time period was equal to **$33,023.00**. The correlation coefficient measurement for these two variables is equal to 0.56317 (rounded to 5 decimal places), this means that 56.317% of the reason for which one variable moves in a specific direction can be explained by the variation in the other variable i.e. based upon the statistics provided 56.317% of the variation in the level of income per capita in the state of **Illinois** at that time period could be explained by the variation in the level of tele-accessibility, or access to telecommunications.

The state of **California** had the seventh highest level of tele-accessibility for the year 2001, at a rate of 1.08945, a measurement which was rounded off to five decimal places. This means that on average 108.945% of the residents of **California** at that time period had access to various forms of telecommunication devices, i.e. fixed and mobile phone line services. The level of income per capita for that same time period was equal to **$32,702.00**. The correlation coefficient measurement for these two variables is equal to 0.56317 (rounded to 5 decimal places), this means that 56.317% of the reason for which one variable moves in a specific direction can be explained by the variation in the other variable i.e. based upon the statistics provided 56.317% of the variation in the level of income per capita in the state of **California** at that time period could be explained by the variation in the level of tele-accessibility, or access to telecommunications.

The state of **Virginia** had the eighth highest level of tele-accessibility for the year 2001, at a rate of 1.08739, a measurement which was rounded off to five decimal places. This means that on average 108.739% of the residents of **Virginia** at that time period had access to various forms of telecommunication devices, i.e. fixed and mobile phone line services. The level of income per capita for that same time period was equal to **$32,431.00**. The correlation coefficient measurement for these two variables is equal to 0.56317 (rounded to 5 decimal places), this means that 56.317% of the reason for which one variable moves in a specific direction can be explained by the variation in the other variable i.e. based upon the statistics provided 56.317% of the variation in the level of income per capita in the state of **Virginia** at that time period could be explained by the variation in the level of tele-accessibility, or access to telecommunications.

The state of **New York** had the ninth highest level of tele-accessibility for the year 2001, at a rate of 1.03859, a measurement which was rounded off to five decimal places. This means that on average 103.859% of the residents of **New York** at that time period had access to various forms of telecommunication devices, i.e. fixed and mobile phone line services. The level of income per capita for that same time period was equal to **$36,019.00.** The correlation coefficient measurement for these two variables is equal to 0.56317 (rounded to 5 decimal places), this means that 56.317% of the reason for which one variable moves in a specific direction can be explained by the variation in the other variable i.e. based upon the statistics provided 56.317% of the variation in the level of income per capita in the state of **New York** at that time period could be explained by the variation in the level of tele-accessibility, or access to telecommunications.

The state of **Pennsylvania** had the tenth highest level of tele-accessibility for the year 2001, at a rate of 1.03090, a measurement which was rounded off to five decimal places. This means that on average 103.090% of the residents of **Pennsylvania** at that time period had access to various forms of telecommunication devices, i.e. fixed and mobile phone line services. The level of income per capita for that same time period was equal to **$30,720.00.** The correlation coefficient measurement for these two variables is equal to 0.56317 (rounded to 5 decimal places), this means that 56.317% of the reason for which one variable moves in a specific direction can be explained by the variation in the other variable i.e. based upon the statistics provided 56.317% of the variation in the level of income per capita in the state of **Pennsylvania** at that time period could be explained by the variation in the level of tele-accessibility, or access to telecommunications.

The state of **Michigan** had the eleventh highest level of tele-accessibility for the year 2001, at a rate of 1.02142, a measurement which was rounded off to five decimal places. This means that on average 102.142% of the residents of **Michigan** at that time period had access to various forms of telecommunication devices, i.e. fixed and mobile phone line services. The level of income per capita for that same time period was equal to **$29,788.00**. The correlation coefficient measurement for these two variables is equal to 0.56317 (rounded to 5 decimal places), this means that 56.317% of the reason for which one variable moves in a specific direction can be explained by the variation in the other variable i.e. based upon the statistics provided 56.317% of the variation in the level of income per capita in the state of **Michigan** at that time period could be explained by the variation in the level of tele-accessibility, or access to telecommunications.

The state of **Ohio** had the twelfth highest level of tele-accessibility for the year 2001, at a rate of 0.99234, a measurement which was rounded off to five decimal places. This means that on average 99.234% of the residents of **Ohio** at that time period had access to various forms of telecommunication devices, i.e. fixed and mobile phone line services. The level of income per capita for that same time period was equal to **$28,816.00**. The

correlation coefficient measurement for these two variables is equal to 0.56317 (rounded to 5 decimal places), this means that 56.317% of the reason for which one variable moves in a specific direction can be explained by the variation in the other variable i.e. based upon the statistics provided 56.317% of the variation in the level of income per capita in the state of **Ohio** at that time period could be explained by the variation in the level of tele-accessibility, or access to telecommunications.

The state of **Tennessee** had the thirteenth highest level of tele-accessibility for the year 2001, at a rate of 0.97945, a measurement which was rounded off to five decimal places. This means that on average 97.945% of the residents of **Tennessee** at that time period had access to various forms of telecommunication devices, i.e. fixed and mobile phone line services. The level of income per capita for that same time period was equal to **$26,988.00**. The correlation coefficient measurement for these two variables is equal to 0.56317 (rounded to 5 decimal places), this means that 56.317% of the reason for which one variable moves in a specific direction can be explained by the variation in the other variable i.e. based upon the statistics provided 56.317% of the variation in the level of income per capita in the state of **Tennessee** at that time period could be explained by the variation in the level of tele-accessibility, or access to telecommunications.

The state of **Indiana** had the fourteenth highest level of tele-accessibility for the year 2001, at a rate of 0.91182, a measurement which was rounded off to five decimal places. This means that on average 91.182% of the residents of **Indiana** at that time period had access to various forms of telecommunication devices, i.e. fixed and mobile phone line services. The level of income per capita for that same time period was equal to **$27,783.00**. The correlation coefficient measurement for these two variables is equal to 0.56317 (rounded to 5 decimal places), this means that 56.317% of the reason for which one variable moves in a specific direction can be explained by the variation in the other variable i.e. based upon the statistics provided 56.317% of the variation in the level of income per capita in the state of **Indiana** at that time period could be explained by the variation in the level of tele-accessibility, or access to telecommunications.

The state of **Utah** had the lowest level of tele-accessibility for the year 2001, at a rate of 0.87548, a measurement which was rounded off to five decimal places. This means that on average 87.548% of the residents of **Utah** at that time period had access to various forms of telecommunication devices, i.e. fixed and mobile phone line services. The level of income per capita for that same time period was equal to **$24,180.00**. The correlation coefficient measurement for these two variables is equal to 0.56317 (rounded to 5 decimal places), this means that 56.317% of the reason for which one variable moves in a specific direction can be explained by the variation in the other variable i.e. based upon the statistics provided 56.317% of the variation in the level of income per capita in the

state of **Utah** at that time period could be explained by the variation in the level of tele-

accessibility, or access to telecommunications.

The level of Tele-Accessibility in the year 2002

The state of **District of Columbia** had the highest level of tele-accessibility for the year 2002, at a rate of 2.42845, a measurement which was rounded off to five decimal places. This means that on average 242.845% of the residents of **District of Columbia** at that time period had access to various forms of telecommunication devices, i.e. fixed and mobile phone line services. The level of income per capita for that same time period was equal to **$48,342.00**. The correlation coefficient measurement for these two variables is equal to 0.74783 (rounded to 5 decimal places), this means that 74.783% of the reason for which one variable moves in a specific direction can be explained by the variation in the other variable i.e. based upon the statistics provided 74.783% of the variation in the level of income per capita in the state of **District of Columbia** at that time period could be explained by the variation in the level of tele-accessibility, or access to telecommunications.

The state of **New Jersey** had the second highest level of tele-accessibility for the year 2002, at a rate of 1.30051, a measurement which was rounded off to five decimal places. This means that on average 130.051% of the residents of **New Jersey** at that time period had access to various forms of telecommunication devices, i.e. fixed and mobile phone line services. The level of income per capita for that same time period was equal to

$40,427.00. The correlation coefficient measurement for these two variables is equal to 0.74783 (rounded to 5 decimal places), this means that 74.783% of the reason for which one variable moves in a specific direction can be explained by the variation in the other variable i.e. based upon the statistics provided 74.783% of the variation in the level of income per capita in the state of **New Jersey** at that time period could be explained by the variation in the level of tele-accessibility, or access to telecommunications

The state of **Florida** had the third highest level of tele-accessibility for the year 2002, at a rate of 1.22953, a measurement which was rounded off to five decimal places. This means that on average 122.953% of the residents of **Florida** at that time period had access to various forms of telecommunication devices, i.e. fixed and mobile phone line services. The level of income per capita for that same time period was equal to **$30,446.00**. The correlation coefficient measurement for these two variables is equal to 0.74783 (rounded to 5 decimal places), this means that 74.783% of the reason for which one variable moves in a specific direction can be explained by the variation in the other variable i.e. based upon the statistics provided 74.783% of the variation in the level of income per capita in the state of **Florida** at that time period could be explained by the variation in the level of tele-accessibility, or access to telecommunications

The state of **Massachusetts** had the fourth highest level of tele-accessibility for the year 2002, at a rate of 1.20966, a measurement which was rounded off to five decimal places. This means that on average 120.966% of the residents of **Massachusetts** at that time period had access to various forms of telecommunication devices, i.e. fixed and mobile phone line services. The level of income per capita for that same time period was equal to **$39,815.00**. The correlation coefficient measurement for these two variables is equal to 0.74783 (rounded to 5 decimal places), this means that 74.783% of the reason for which one variable moves in a specific direction can be explained by the variation in the other variable i.e. based upon the statistics provided 74.783% of the variation in the level of income per capita in the state of **Massachusetts** at that time period could be explained by the variation in the level of tele-accessibility, or access to telecommunications.

The state of **Connecticut** had the fifth highest level of tele-accessibility for the year 2002, at a rate of 1.18252, a measurement which was rounded off to five decimal places. This means that on average 118.252% of the residents of **Connecticut** at that time period had access to various forms of telecommunication devices, i.e. fixed and mobile phone line services. The level of income per capita for that same time period was equal to **$43,173.00**. The correlation coefficient measurement for these two variables is equal to 0.74783 (rounded to 5 decimal places), this means that 74.783% of the reason for which

one variable moves in a specific direction can be explained by the variation in the other variable i.e. based upon the statistics provided 74.783% of the variation in the level of income per capita in the state of **Connecticut** at that time period could be explained by the variation in the level of tele-accessibility, or access to telecommunications.

The state of **California** had the sixth highest level of tele-accessibility for the year 2002, at a rate of 1.15213, a measurement which was rounded off to five decimal places. This means that on average 115.213% of the residents of **California** at that time period had access to various forms of telecommunication devices, i.e. fixed and mobile phone line services. The level of income per capita for that same time period was equal to **$33,749.00**. The correlation coefficient measurement for these two variables is equal to 0.74783 (rounded to 5 decimal places), this means that 74.783% of the reason for which one variable moves in a specific direction can be explained by the variation in the other variable i.e. based upon the statistics provided 74.783% of the variation in the level of income per capita in the state of **California** at that time period could be explained by the variation in the level of tele-accessibility, or access to telecommunications.

The state of **Virginia** had the seventh highest level of tele-accessibility for the year 2002, at a rate of 1.14389, a measurement which was rounded off to five decimal places. This means that on average 114.389% of the residents of **Virginia** at that time period had access to various forms of telecommunication devices, i.e. fixed and mobile phone line services. The level of income per capita for that same time period was equal to **$33,671.00**. The correlation coefficient measurement for these two variables is equal to 0.74783 (rounded to 5 decimal places), this means that 74.783% of the reason for which one variable moves in a specific direction can be explained by the variation in the other variable i.e. based upon the statistics provided 74.783% of the variation in the level of income per capita in the state of **Virginia** at that time period could be explained by the variation in the level of tele-accessibility, or access to telecommunications.

The state of **Michigan** had the eighth highest level of tele-accessibility for the year 2002, at a rate of 1.12516, a measurement which was rounded off to five decimal places. This means that on average 112.516% of the residents of **Michigan** at that time period had access to various forms of telecommunication devices, i.e. fixed and mobile phone line services. The level of income per capita for that same time period was equal to **$30,439.00**. The correlation coefficient measurement for these two variables is equal to 0.74783 (rounded to 5 decimal places), this means that 74.783% of the reason for which one variable moves in a specific direction can be explained by the variation in the other variable i.e. based upon the statistics provided 74.783% of the variation in the level of income per capita in the state of **Michigan** at that time period could be explained by the variation in the level of tele-accessibility, or access to telecommunications.

The state of **Illinois** had the ninth highest level of tele-accessibility for the year 2002, at a rate of 1.11528, a measurement which was rounded off to five decimal places. This means that on average 111.528% of the residents of **Illinois** at that time period had access to various forms of telecommunication devices, i.e. fixed and mobile phone line services. The level of income per capita for that same time period was equal to **$33,690.00**. The correlation coefficient measurement for these two variables is equal to 0.74783 (rounded to 5 decimal places), this means that 74.783% of the reason for which one variable moves in a specific direction can be explained by the variation in the other variable i.e. based upon the statistics provided 74.783% of the variation in the level of income per capita in the state of **Illinois** at that time period could be explained by the variation in the level of tele-accessibility, or access to telecommunications.

The state of **Pennsylvania** had the tenth highest level of tele-accessibility for the year 2002, at a rate of 1.10010, a measurement which was rounded off to five decimal places. This means that on average 110.010% of the residents of **Pennsylvania** at that time period had access to various forms of telecommunication devices, i.e. fixed and mobile phone line services. The level of income per capita for that same time period was equal to **$31,998.00**. The correlation coefficient measurement for these two variables is equal to 0.74783 (rounded to 5 decimal places), this means that 74.783% of the reason for which one variable moves in a specific direction can be explained by the variation in the other variable i.e. based upon the statistics provided 74.783% of the variation in the level of income per capita in the state of **Pennsylvania** at that time period could be explained by the variation in the level of tele-accessibility, or access to telecommunications.

The state of **New York** had the eleventh highest level of tele-accessibility for the year 2002, at a rate of 1.08219, a measurement which was rounded off to five decimal places. This means that on average 108.219% of the residents of **New York** at that time period had access to various forms of telecommunication devices, i.e. fixed and mobile phone line services. The level of income per capita for that same time period was equal to **$36,574.00**. The correlation coefficient measurement for these two variables is equal to 0.74783 (rounded to 5 decimal places), this means that 74.783% of the reason for which one variable moves in a specific direction can be explained by the variation in the other variable i.e. based upon the statistics provided 74.783% of the variation in the level of income per capita in the state of **New York** at that time period could be explained by the variation in the level of tele-accessibility, or access to telecommunications.

The state of **Tennessee** had the twelfth highest level of tele-accessibility for the year 2002, at a rate of 1.05703, a measurement which was rounded off to five decimal places. This means that on average 105.703% of the residents of **Tennessee** at that time period had access to various forms of telecommunication devices, i.e. fixed and mobile phone line services. The level of income per capita for that same time period was equal to

$28,455.00. The correlation coefficient measurement for these two variables is equal to 0.74783 (rounded to 5 decimal places), this means that 74.783% of the reason for which one variable moves in a specific direction can be explained by the variation in the other variable i.e. based upon the statistics provided 74.783% of the variation in the level of income per capita in the state of **Tennessee** at that time period could be explained by the variation in the level of tele-accessibility, or access to telecommunications.

The state of **Ohio** had the thirteenth highest level of tele-accessibility for the year 2002, at a rate of 1.04589, a measurement which was rounded off to five decimal places. This means that on average 104.589% of the residents of **Ohio** at that time period had access to various forms of telecommunication devices, i.e. fixed and mobile phone line services. The level of income per capita for that same time period was equal to **$29,944.00**. The correlation coefficient measurement for these two variables is equal to 0.74783 (rounded to 5 decimal places), this means that 74.783% of the reason for which one variable moves in a specific direction can be explained by the variation in the other variable i.e. based upon the statistics provided 74.783% of the variation in the level of income per capita in the state of **Ohio** at that time period could be explained by the variation in the level of tele-accessibility, or access to telecommunications.

The state of **Utah** had the fourteenth highest level of tele-accessibility for the year 2002, at a rate of 0.95965, a measurement which was rounded off to five decimal places. This means that on average 95.965% of the residents of **Utah** at that time period had access to various forms of telecommunication devices, i.e. fixed and mobile phone line services. The level of income per capita for that same time period was equal to **$24,977.00**. The correlation coefficient measurement for these two variables is equal to 0.74783 (rounded to 5 decimal places), this means that 74.783% of the reason for which one variable moves in a specific direction can be explained by the variation in the other variable i.e. based upon the statistics provided 74.783% of the variation in the level of income per capita in the state of **Utah** at that time period could be explained by the variation in the level of tele-accessibility, or access to telecommunications.

The state of **Indiana** had the lowest level of tele-accessibility for the year 2002, at a rate of 0.93945, a measurement which was rounded off to five decimal places. This means that on average 93.945% of the residents of **Indiana** at that time period had access to various forms of telecommunication devices, i.e. fixed and mobile phone line services. The level of income per capita for that same time period was equal to **$28,783.00**. The correlation coefficient measurement for these two variables is equal to 0.74783 (rounded to 5 decimal places), this means that 74.783% of the reason for which one variable moves in a specific direction can be explained by the variation in the other variable i.e. based upon the statistics provided 74.783% of the variation in the level of income per capita in the state of **Indiana** at that time period could be explained by the variation in

the level of tele-accessibility, or access to telecommunications.

The level of Tele-Accessibility in the year 2003

The state of **District of Columbia** had the highest level of tele-accessibility for the year 2003, at a rate of 2.53965, a measurement which was rounded off to five decimal places. This means that on average 253.965% of the residents of **District of Columbia** at that time period had access to various forms of telecommunication devices, i.e. fixed and mobile phone line services. The level of income per capita for that same time period was equal to **$47,529.00**. The correlation coefficient measurement for these two variables is equal to 0.73257 (rounded to 5 decimal places), this means that 73.257% of the reason for which one variable moves in a specific direction can be explained by the variation in the other variable i.e. based upon the statistics provided 73.257% of the variation in the level of income per capita in the state of **District of Columbia** at that time period could be explained by the variation in the level of tele-accessibility, or access to telecommunications.

The state of **New Jersey** had the second highest level of tele-accessibility for the year 2003, at a rate of 1.37379, a measurement which was rounded off to five decimal places. This means that on average 137.379% of the residents of **New Jersey** at that time period had access to various forms of telecommunication devices, i.e. fixed and mobile phone line services. The level of income per capita for that same time period was equal to

$40,504.00. The correlation coefficient measurement for these two variables is equal to 0.73257 (rounded to 5 decimal places), this means that 73.257% of the reason for which one variable moves in a specific direction can be explained by the variation in the other variable i.e. based upon the statistics provided 73.257% of the variation in the level of income per capita in the state of **New Jersey** at that time period could be explained by the variation in the level of tele-accessibility, or access to telecommunications

The state of **Florida** had the third highest level of tele-accessibility for the year 2003, at a rate of 1.29107, a measurement which was rounded off to five decimal places. This means that on average 129.107% of the residents of **Florida** at that time period had access to various forms of telecommunication devices, i.e. fixed and mobile phone line services. The level of income per capita for that same time period was equal to **$31,364.00**. The correlation coefficient measurement for these two variables is equal to 0.73257 (rounded to 5 decimal places), this means that 73.257% of the reason for which one variable moves in a specific direction can be explained by the variation in the other variable i.e. based upon the statistics provided 73.257% of the variation in the level of income per capita in the state of **Florida** at that time period could be explained by the variation in the level of tele-accessibility, or access to telecommunications

The state of **Massachusetts** had the fourth highest level of tele-accessibility for the year 2003, at a rate of 1.22667, a measurement which was rounded off to five decimal places. This means that on average 122.667% of the residents of **Massachusetts** at that time period had access to various forms of telecommunication devices, i.e. fixed and mobile phone line services. The level of income per capita for that same time period was equal to **$40,161.00**. The correlation coefficient measurement for these two variables is equal to 0.73257 (rounded to 5 decimal places), this means that 73.257% of the reason for which one variable moves in a specific direction can be explained by the variation in the other variable i.e. based upon the statistics provided 73.257% of the variation in the level of income per capita in the state of **Massachusetts** at that time period could be explained by the variation in the level of tele-accessibility, or access to telecommunications.

The state of **Connecticut** had the fifth highest level of tele-accessibility for the year 2003, at a rate of 1.22326, a measurement which was rounded off to five decimal places. This means that on average 122.326% of the residents of **Connecticut** at that time period had access to various forms of telecommunication devices, i.e. fixed and mobile phone line services. The level of income per capita for that same time period was equal to **$43,730.00.** The correlation coefficient measurement for these two variables is equal to 0.73257 (rounded to 5 decimal places), this means that 73.257% of the reason for which

one variable moves in a specific direction can be explained by the variation in the other variable i.e. based upon the statistics provided 73.257% of the variation in the level of income per capita in the state of **Connecticut** at that time period could be explained by the variation in the level of tele-accessibility, or access to telecommunications.

The state of **California** had the sixth highest level of tele-accessibility for the year 2003, at a rate of 1.20805, a measurement which was rounded off to five decimal places. This means that on average 120.805% of the residents of **California** at that time period had access to various forms of telecommunication devices, i.e. fixed and mobile phone line services. The level of income per capita for that same time period was equal to **$34,922.00**. The correlation coefficient measurement for these two variables is equal to 0.73257 (rounded to 5 decimal places), this means that 73.257% of the reason for which one variable moves in a specific direction can be explained by the variation in the other variable i.e. based upon the statistics provided 73.257% of the variation in the level of income per capita in the state of **California** at that time period could be explained by the variation in the level of tele-accessibility, or access to telecommunications.

The state of **Illinois** had the seventh highest level of tele-accessibility for the year 2003, at a rate of 1.20592, a measurement which was rounded off to five decimal places. This means that on average 120.592% of the residents of **Illinois** at that time period had access to various forms of telecommunication devices, i.e. fixed and mobile phone line services. The level of income per capita for that same time period was equal to **$34,569.00**. The correlation coefficient measurement for these two variables is equal to 0.73257 (rounded to 5 decimal places), this means that 73.257% of the reason for which one variable moves in a specific direction can be explained by the variation in the other variable i.e. based upon the statistics provided 73.257% of the variation in the level of income per capita in the state of **Illinois** at that time period could be explained by the variation in the level of tele-accessibility, or access to telecommunications.

The state of **Virginia** had the eighth highest level of tele-accessibility for the year 2003, at a rate of 1.17161, a measurement which was rounded off to five decimal places. This means that on average 117.161% of the residents of **Virginia** at that time period had access to various forms of telecommunication devices, i.e. fixed and mobile phone line services. The level of income per capita for that same time period was equal to **$35,029.00**. The correlation coefficient measurement for these two variables is equal to 0.73257 (rounded to 5 decimal places), this means that 73.257% of the reason for which one variable moves in a specific direction can be explained by the variation in the other variable i.e. based upon the statistics provided 73.257% of the variation in the level of income per capita in the state of **Virginia** at that time period could be explained by the variation in the level of tele-accessibility, or access to telecommunications.

The state of **Pennsylvania** had the ninth highest level of tele-accessibility for the year 2003, at a rate of 1.12832, a measurement which was rounded off to five decimal places. This means that on average 112.832% of the residents of **Pennsylvania** at that time period had access to various forms of telecommunication devices, i.e. fixed and mobile phone line services. The level of income per capita for that same time period was equal to **$32,427.00**. The correlation coefficient measurement for these two variables is equal to 0.73257 (rounded to 5 decimal places), this means that 73.257% of the reason for which one variable moves in a specific direction can be explained by the variation in the other variable i.e. based upon the statistics provided 73.257% of the variation in the level of income per capita in the state of **Pennsylvania** at that time period could be explained by the variation in the level of tele-accessibility, or access to telecommunications.

The state of **New York** had the tenth highest level of tele-accessibility for the year 2003, at a rate of 1.10900, a measurement which was rounded off to five decimal places. This means that on average 110.900% of the residents of **New York** at that time period had access to various forms of telecommunication devices, i.e. fixed and mobile phone line services. The level of income per capita for that same time period was equal to **$36,165.00**. The correlation coefficient measurement for these two variables is equal to 0.73257 (rounded to 5 decimal places), this means that 73.257% of the reason for which one variable moves in a specific direction can be explained by the variation in the other variable i.e. based upon the statistics provided 73.257% of the variation in the level of

income per capita in the state of **New York** at that time period could be explained by the variation in the level of tele-accessibility, or access to telecommunications.

The state of **Michigan** had the eleventh highest level of tele-accessibility for the year 2003, at a rate of 1.10204, a measurement which was rounded off to five decimal places. This means that on average 110.204% of the residents of **Michigan** at that time period had access to various forms of telecommunication devices, i.e. fixed and mobile phone line services. The level of income per capita for that same time period was equal to **$31,214.00**. The correlation coefficient measurement for these two variables is equal to 0.73257 (rounded to 5 decimal places), this means that 73.257% of the reason for which one variable moves in a specific direction can be explained by the variation in the other variable i.e. based upon the statistics provided 73.257% of the variation in the level of income per capita in the state of **Michigan** at that time period could be explained by the variation in the level of tele-accessibility, or access to telecommunications.

The state of **Ohio** had the twelfth highest level of tele-accessibility for the year 2003, at a rate of 1.09612, a measurement which was rounded off to five decimal places. This means that on average 109.612% of the residents of **Ohio** at that time period had access to various forms of telecommunication devices, i.e. fixed and mobile phone line services. The level of income per capita for that same time period was equal to **$30,698.00**. The

correlation coefficient measurement for these two variables is equal to 0.73257 (rounded to 5 decimal places), this means that 73.257% of the reason for which one variable moves in a specific direction can be explained by the variation in the other variable i.e. based upon the statistics provided 73.257% of the variation in the level of income per capita in the state of **Ohio** at that time period could be explained by the variation in the level of tele-accessibility, or access to telecommunications.

The state of **Tennessee** had the thirteenth highest level of tele-accessibility for the year 2003, at a rate of 1.05686, a measurement which was rounded off to five decimal places. This means that on average 105.686% of the residents of **Tennessee** at that time period had access to various forms of telecommunication devices, i.e. fixed and mobile phone line services. The level of income per capita for that same time period was equal to **$29,026.00**. The correlation coefficient measurement for these two variables is equal to 0.73257 (rounded to 5 decimal places), this means that 73.257% of the reason for which one variable moves in a specific direction can be explained by the variation in the other variable i.e. based upon the statistics provided 73.257% of the variation in the level of income per capita in the state of **Tennessee** at that time period could be explained by the variation in the level of tele-accessibility, or access to telecommunications.

The state of **Indiana** had the fourteenth highest level of tele-accessibility for the year 2003, at a rate of 0.99193, a measurement which was rounded off to five decimal places. This means that on average 99.193% of the residents of **Indiana** at that time period had access to various forms of telecommunication devices, i.e. fixed and mobile phone line services. The level of income per capita for that same time period was equal to **$29,588.00**. The correlation coefficient measurement for these two variables is equal to 0.73257 (rounded to 5 decimal places), this means that 73.257% of the reason for which one variable moves in a specific direction can be explained by the variation in the other variable i.e. based upon the statistics provided 73.257% of the variation in the level of income per capita in the state of **Indiana** at that time period could be explained by the variation in the level of tele-accessibility, or access to telecommunications.

The state of **Utah** had the lowest level of tele-accessibility for the year 2003, at a rate of 0.98693, a measurement which was rounded off to five decimal places. This means that on average 98.693% of the residents of **Utah** at that time period had access to various forms of telecommunication devices, i.e. fixed and mobile phone line services. The level of income per capita for that same time period was equal to **$25,830.00**. The correlation coefficient measurement for these two variables is equal to 0.73257 (rounded to 5 decimal places), this means that 73.257% of the reason for which one variable moves in a specific direction can be explained by the variation in the other variable i.e. based upon the statistics provided 73.257% of the variation in the level of income per capita in the

state of **Utah** at that time period could be explained by the variation in the level of tele-accessibility, or access to telecommunications.

The level of Tele-Accessibility in the year 2004

The state of **District of Columbia** had the highest level of tele-accessibility for the year 2004, at a rate of 2.90958, a measurement which was rounded off to five decimal places. This means that on average 290.958% of the residents of **District of Columbia** at that time period had access to various forms of telecommunication devices, i.e. fixed and mobile phone line services. The level of income per capita for that same time period was equal to **$51,458.00**. The correlation coefficient measurement for these two variables is equal to 0.75384 (rounded to 5 decimal places), this means that 75.384% of the reason for which one variable moves in a specific direction can be explained by the variation in the other variable i.e. based upon the statistics provided 75.384% of the variation in the level of income per capita in the state of **District of Columbia** at that time period could be explained by the variation in the level of tele-accessibility, or access to telecommunications.

The state of **New Jersey** had the second highest level of tele-accessibility for the year 2004, at a rate of 1.48575, a measurement which was rounded off to five decimal places. This means that on average 148.575% of the residents of **New Jersey** at that time period had access to various forms of telecommunication devices, i.e. fixed and mobile phone line services. The level of income per capita for that same time period was equal to

$42,406.00. The correlation coefficient measurement for these two variables is equal to 0.75384 (rounded to 5 decimal places), this means that 75.384% of the reason for which one variable moves in a specific direction can be explained by the variation in the other variable i.e. based upon the statistics provided 75.384% of the variation in the level of income per capita in the state of **New Jersey** at that time period could be explained by the variation in the level of tele-accessibility, or access to telecommunications

The state of **Florida** had the third highest level of tele-accessibility for the year 2004, at a rate of 1.34301, a measurement which was rounded off to five decimal places. This means that on average 134.301% of the residents of **Florida** at that time period had access to various forms of telecommunication devices, i.e. fixed and mobile phone line services. The level of income per capita for that same time period was equal to **$33,659.00**. The correlation coefficient measurement for these two variables is equal to 0.75384 (rounded to 5 decimal places), this means that 75.384% of the reason for which one variable moves in a specific direction can be explained by the variation in the other variable i.e. based upon the statistics provided 75.384% of the variation in the level of income per capita in the state of **Florida** at that time period could be explained by the variation in the level of tele-accessibility, or access to telecommunications

The state of **Massachusetts** had the fourth highest level of tele-accessibility for the year 2004, at a rate of 1.29415, a measurement which was rounded off to five decimal places. This means that on average 129.415% of the residents of **Massachusetts** at that time period had access to various forms of telecommunication devices, i.e. fixed and mobile phone line services. The level of income per capita for that same time period was equal to **$42,123.00**. The correlation coefficient measurement for these two variables is equal to 0.75384 (rounded to 5 decimal places), this means that 75.384% of the reason for which one variable moves in a specific direction can be explained by the variation in the other variable i.e. based upon the statistics provided 75.384% of the variation in the level of income per capita in the state of **Massachusetts** at that time period could be explained by the variation in the level of tele-accessibility, or access to telecommunications.

The state of **Connecticut** had the fifth highest level of tele-accessibility for the year 2004, at a rate of 1.27763, a measurement which was rounded off to five decimal places. This means that on average 127.763% of the residents of **Connecticut** at that time period had access to various forms of telecommunication devices, i.e. fixed and mobile phone line services. The level of income per capita for that same time period was equal to **$46,417.00**. The correlation coefficient measurement for these two variables is equal to 0.75384 (rounded to 5 decimal places), this means that 75.384% of the reason for which

one variable moves in a specific direction can be explained by the variation in the other variable i.e. based upon the statistics provided 75.384% of the variation in the level of income per capita in the state of **Connecticut** at that time period could be explained by the variation in the level of tele-accessibility, or access to telecommunications.

The state of **Virginia** had the sixth highest level of tele-accessibility for the year 2004, at a rate of 1.26688, a measurement which was rounded off to five decimal places. This means that on average 126.688% of the residents of **Virginia** at that time period had access to various forms of telecommunication devices, i.e. fixed and mobile phone line services. The level of income per capita for that same time period was equal to **$36,912.00**. The correlation coefficient measurement for these two variables is equal to 0.75384 (rounded to 5 decimal places), this means that 75.384% of the reason for which one variable moves in a specific direction can be explained by the variation in the other variable i.e. based upon the statistics provided 75.384% of the variation in the level of income per capita in the state of **Virginia** at that time period could be explained by the variation in the level of tele-accessibility, or access to telecommunications.

The state of **California** had the seventh highest level of tele-accessibility for the year 2004, at a rate of 1.25929, a measurement which was rounded off to five decimal places.

This means that on average 125.929% of the residents of **California** at that time period had access to various forms of telecommunication devices, i.e. fixed and mobile phone line services. The level of income per capita for that same time period was equal to **$36,830.00**. The correlation coefficient measurement for these two variables is equal to 0.75384 (rounded to 5 decimal places), this means that 75.384% of the reason for which one variable moves in a specific direction can be explained by the variation in the other variable i.e. based upon the statistics provided 75.384% of the variation in the level of income per capita in the state of **California** at that time period could be explained by the variation in the level of tele-accessibility, or access to telecommunications.

The state of **Illinois** had the eighth highest level of tele-accessibility for the year 2004, at a rate of 1.22808, a measurement which was rounded off to five decimal places. This means that on average 122.808% of the residents of **Illinois** at that time period had access to various forms of telecommunication devices, i.e. fixed and mobile phone line services. The level of income per capita for that same time period was equal to **$35,957.00**. The correlation coefficient measurement for these two variables is equal to 0.75384 (rounded to 5 decimal places), this means that 75.384% of the reason for which one variable moves in a specific direction can be explained by the variation in the other variable i.e. based upon the statistics provided 75.384% of the variation in the level of income per capita in the state of **Illinois** at that time period could be explained by the variation in the level of tele-accessibility, or access to telecommunications.

The state of **Pennsylvania** had the ninth highest level of tele-accessibility for the year 2004, at a rate of 1.19185, a measurement which was rounded off to five decimal places. This means that on average 119.185% of the residents of **Pennsylvania** at that time period had access to various forms of telecommunication devices, i.e. fixed and mobile phone line services. The level of income per capita for that same time period was equal to **$33,852.00**. The correlation coefficient measurement for these two variables is equal to 0.75384 (rounded to 5 decimal places), this means that 75.384% of the reason for which one variable moves in a specific direction can be explained by the variation in the other variable i.e. based upon the statistics provided 75.384% of the variation in the level of income per capita in the state of **Pennsylvania** at that time period could be explained by the variation in the level of tele-accessibility, or access to telecommunications.

The state of **New York** had the tenth highest level of tele-accessibility for the year 2004, at a rate of 1.15606, a measurement which was rounded off to five decimal places. This means that on average 115.606% of the residents of **New York** at that time period had access to various forms of telecommunication devices, i.e. fixed and mobile phone line services. The level of income per capita for that same time period was equal to **$38,398.00**. The correlation coefficient measurement for these two variables is equal to 0.75384 (rounded to 5 decimal places), this means that 75.384% of the reason for which one variable moves in a specific direction can be explained by the variation in the other variable i.e. based upon the statistics provided 75.384% of the variation in the level of

income per capita in the state of **New York** at that time period could be explained by the variation in the level of tele-accessibility, or access to telecommunications.

The state of **Michigan** had the eleventh highest level of tele-accessibility for the year 2004, at a rate of 1.13918, a measurement which was rounded off to five decimal places. This means that on average 113.918% of the residents of **Michigan** at that time period had access to various forms of telecommunication devices, i.e. fixed and mobile phone line services. The level of income per capita for that same time period was equal to **$31,650.00**. The correlation coefficient measurement for these two variables is equal to 0.75384 (rounded to 5 decimal places), this means that 75.384% of the reason for which one variable moves in a specific direction can be explained by the variation in the other variable i.e. based upon the statistics provided 75.384% of the variation in the level of income per capita in the state of **Michigan** at that time period could be explained by the variation in the level of tele-accessibility, or access to telecommunications.

The state of **Ohio** had the twelfth highest level of tele-accessibility for the year 2004, at a rate of 1.12218, a measurement which was rounded off to five decimal places. This means that on average 112.218% of the residents of **Ohio** at that time period had access to various forms of telecommunication devices, i.e. fixed and mobile phone line services. The level of income per capita for that same time period was equal to **$31,617.00**. The

correlation coefficient measurement for these two variables is equal to 0.75384 (rounded to 5 decimal places), this means that 75.384% of the reason for which one variable moves in a specific direction can be explained by the variation in the other variable i.e. based upon the statistics provided 75.384% of the variation in the level of income per capita in the state of **Ohio** at that time period could be explained by the variation in the level of tele-accessibility, or access to telecommunications.

The state of **Tennessee** had the thirteenth highest level of tele-accessibility for the year 2004, at a rate of 1.09275, a measurement which was rounded off to five decimal places. This means that on average 109.275% of the residents of **Tennessee** at that time period had access to various forms of telecommunication devices, i.e. fixed and mobile phone line services. The level of income per capita for that same time period was equal to **$30,297.00**. The correlation coefficient measurement for these two variables is equal to 0.75384 (rounded to 5 decimal places), this means that 75.384% of the reason for which one variable moves in a specific direction can be explained by the variation in the other variable i.e. based upon the statistics provided 75.384% of the variation in the level of income per capita in the state of **Tennessee** at that time period could be explained by the variation in the level of tele-accessibility, or access to telecommunications.

The state of **Indiana** had the fourteenth highest level of tele-accessibility for the year 2004, at a rate of 1.03654, a measurement which was rounded off to five decimal places. This means that on average 103.654% of the residents of **Indiana** at that time period had access to various forms of telecommunication devices, i.e. fixed and mobile phone line services. The level of income per capita for that same time period was equal to **$30,645.00**. The correlation coefficient measurement for these two variables is equal to 0.75384 (rounded to 5 decimal places), this means that 75.384% of the reason for which one variable moves in a specific direction can be explained by the variation in the other variable i.e. based upon the statistics provided 75.384% of the variation in the level of income per capita in the state of **Indiana** at that time period could be explained by the variation in the level of tele-accessibility, or access to telecommunications.

The state of **Utah** had the lowest level of tele-accessibility for the year 2004, at a rate of 1.00771, a measurement which was rounded off to five decimal places. This means that on average 100.771% of the residents of **Utah** at that time period had access to various forms of telecommunication devices, i.e. fixed and mobile phone line services. The level of income per capita for that same time period was equal to **$26,827.00**. The correlation coefficient measurement for these two variables is equal to 0.75384 (rounded to 5 decimal places), this means that 75.384% of the reason for which one variable moves in a specific direction can be explained by the variation in the other variable i.e. based upon the statistics provided 75.384% of the variation in the level of income per capita in the

state of **Utah** at that time period could be explained by the variation in the level of tele-accessibility, or access to telecommunications.

The level of Tele-Accessibility in the year 2005

The state of **District of Columbia** had the highest level of tele-accessibility for the year 2005, at a rate of 2.65242, a measurement which was rounded off to five decimal places. This means that on average 265.242% of the residents of **District of Columbia** at that time period had access to various forms of telecommunication devices, i.e. fixed and mobile phone line services. The level of income per capita for that same time period was equal to **$55,268.00**. The correlation coefficient measurement for these two variables is equal to 0.77669 (rounded to 5 decimal places), this means that 77.669% of the reason for which one variable moves in a specific direction can be explained by the variation in the other variable i.e. based upon the statistics provided 77.669 % of the variation in the level of income per capita in the state of **District of Columbia** at that time period could be explained by the variation in the level of tele-accessibility, or access to telecommunications.

The state of **New Jersey** had the second highest level of tele-accessibility for the year 2005, at a rate of 1.41699, a measurement which was rounded off to five decimal places. This means that on average 141.699% of the residents of **New Jersey** at that time period had access to various forms of telecommunication devices, i.e. fixed and mobile phone line services. The level of income per capita for that same time period was equal to

$43,994.00. The correlation coefficient measurement for these two variables is equal to 0.77669 (rounded to 5 decimal places), this means that 77.669% of the reason for which one variable moves in a specific direction can be explained by the variation in the other variable i.e. based upon the statistics provided 77.669% of the variation in the level of income per capita in the state of **New Jersey** at that time period could be explained by the variation in the level of tele-accessibility, or access to telecommunications

The state of **Florida** had the third highest level of tele-accessibility for the year 2005, at a rate of 1.29200, a measurement which was rounded off to five decimal places. This means that on average 129.200% of the residents of **Florida** at that time period had access to various forms of telecommunication devices, i.e. fixed and mobile phone line services. The level of income per capita for that same time period was equal to **$35,769.00**. The correlation coefficient measurement for these two variables is equal to 0.77669 (rounded to 5 decimal places), this means that 77.669% of the reason for which one variable moves in a specific direction can be explained by the variation in the other variable i.e. based upon the statistics provided 77.669% of the variation in the level of income per capita in the state of **Florida** at that time period could be explained by the variation in the level of tele-accessibility, or access to telecommunications

The state of **Connecticut** had the fourth highest level of tele-accessibility for the year 2005, at a rate of 1.28371, a measurement which was rounded off to five decimal places. This means that on average 128.371% of the residents of **Connecticut** at that time period had access to various forms of telecommunication devices, i.e. fixed and mobile phone line services. The level of income per capita for that same time period was equal to **$48,485.00**. The correlation coefficient measurement for these two variables is equal to 0.77669 (rounded to 5 decimal places), this means that 77.669% of the reason for which one variable moves in a specific direction can be explained by the variation in the other variable i.e. based upon the statistics provided 77.669% of the variation in the level of income per capita in the state of **Connecticut** at that time period could be explained by the variation in the level of tele-accessibility, or access to telecommunications.

The state of **California** had the fifth highest level of tele-accessibility for the year 2005, at a rate of 1.28109, a measurement which was rounded off to five decimal places. This means that on average 128.109% of the residents of **California** at that time period had access to various forms of telecommunication devices, i.e. fixed and mobile phone line services. The level of income per capita for that same time period was equal to **$38,670.00**. The correlation coefficient measurement for these two variables is equal to 0.77669 (rounded to 5 decimal places), this means that 77.669% of the reason for which one variable moves in a specific direction can be explained by the variation in the other variable i.e. based upon the statistics provided 77.669% of the variation in the level of income per capita in the state of **California** at that time period could be explained by the variation in the level of tele-accessibility, or access to telecommunications.

The state of **Massachusetts** had the sixth highest level of tele-accessibility for the year 2005, at a rate of 1.28107, a measurement which was rounded off to five decimal places. This means that on average 128.107% of the residents of **Massachusetts** at that time period had access to various forms of telecommunication devices, i.e. fixed and mobile phone line services. The level of income per capita for that same time period was equal to **$43,897.00**. The correlation coefficient measurement for these two variables is equal to 0.77669 (rounded to 5 decimal places), this means that 77.669% of the reason for which one variable moves in a specific direction can be explained by the variation in the other variable i.e. based upon the statistics provided 77.669% of the variation in the level of income per capita in the state of **Massachusetts** at that time period could be explained by the variation in the level of tele-accessibility, or access to telecommunications.

The state of **New York** had the seventh highest level of tele-accessibility for the year 2005, at a rate of 1.25602, a measurement which was rounded off to five decimal places.

This means that on average 125.602% of the residents of **New York** at that time period had access to various forms of telecommunication devices, i.e. fixed and mobile phone line services. The level of income per capita for that same time period was equal to **$40,678.00**. The correlation coefficient measurement for these two variables is equal to 0.77669 (rounded to 5 decimal places), this means that 77.669% of the reason for which one variable moves in a specific direction can be explained by the variation in the other variable i.e. based upon the statistics provided 77.669% of the variation in the level of income per capita in the state of **New York** at that time period could be explained by the variation in the level of tele-accessibility, or access to telecommunications.

The state of **Illinois** had the eighth highest level of tele-accessibility for the year 2005, at a rate of 1.22693, a measurement which was rounded off to five decimal places. This means that on average 122.693% of the residents of **Illinois** at that time period had access to various forms of telecommunication devices, i.e. fixed and mobile phone line services. The level of income per capita for that same time period was equal to **$37,168.00**. The correlation coefficient measurement for these two variables is equal to 0.77669 (rounded to 5 decimal places), this means that 77.669% of the reason for which one variable moves in a specific direction can be explained by the variation in the other variable i.e. based upon the statistics provided 77.669% of the variation in the level of income per capita in the state of **Illinois** at that time period could be explained by the variation in the level of tele-accessibility, or access to telecommunications.

The state of **Virginia** had the ninth highest level of tele-accessibility for the year 2005, at a rate of 1.20858, a measurement which was rounded off to five decimal places. This means that on average 120.858% of the residents of **Virginia** at that time period had access to various forms of telecommunication devices, i.e. fixed and mobile phone line services. The level of income per capita for that same time period was equal to **$38,980.00**. The correlation coefficient measurement for these two variables is equal to 0.77669 (rounded to 5 decimal places), this means that 77.669% of the reason for which one variable moves in a specific direction can be explained by the variation in the other variable i.e. based upon the statistics provided 77.669% of the variation in the level of income per capita in the state of **Virginia** at that time period could be explained by the variation in the level of tele-accessibility, or access to telecommunications.

The state of **Tennessee** had the tenth highest level of tele-accessibility for the year 2005, at a rate of 1.19279, a measurement which was rounded off to five decimal places. This means that on average 119.279% of the residents of **Tennessee** at that time period had access to various forms of telecommunication devices, i.e. fixed and mobile phone line services. The level of income per capita for that same time period was equal to **$31,360.00**. The correlation coefficient measurement for these two variables is equal to 0.77669 (rounded to 5 decimal places), this means that 77.669% of the reason for which one variable moves in a specific direction can be explained by the variation in the other variable i.e. based upon the statistics provided 77.669% of the variation in the level of income per capita in the state of **Tennessee** at that time period could be explained by the variation in the level of tele-accessibility, or access to telecommunications.

The state of **Pennsylvania** had the eleventh highest level of tele-accessibility for the year 2005, at a rate of 1.18718, a measurement which was rounded off to five decimal places. This means that on average 118.718% of the residents of **Pennsylvania** at that time period had access to various forms of telecommunication devices, i.e. fixed and mobile phone line services. The level of income per capita for that same time period was equal to **$34,978.00**. The correlation coefficient measurement for these two variables is equal to 0.77669 (rounded to 5 decimal places), this means that 77.669% of the reason for which one variable moves in a specific direction can be explained by the variation in the other variable i.e. based upon the statistics provided 77.669% of the variation in the level of income per capita in the state of **Pennsylvania** at that time period could be explained by the variation in the level of tele-accessibility, or access to telecommunications.

The state of **Michigan** had the twelfth highest level of tele-accessibility for the year 2005, at a rate of 1.18111, a measurement which was rounded off to five decimal places. This means that on average 118.111% of the residents of **Michigan** at that time period had access to various forms of telecommunication devices, i.e. fixed and mobile phone line services. The level of income per capita for that same time period was equal to

$32,265.00. The correlation coefficient measurement for these two variables is equal to 0.77669 (rounded to 5 decimal places), this means that 77.669% of the reason for which one variable moves in a specific direction can be explained by the variation in the other variable i.e. based upon the statistics provided 77.669% of the variation in the level of income per capita in the state of **Michigan** at that time period could be explained by the variation in the level of tele-accessibility, or access to telecommunications.

The state of **Ohio** had the thirteenth highest level of tele-accessibility for the year 2005, at a rate of 1.16476, a measurement which was rounded off to five decimal places. This means that on average 116.476% of the residents of **Ohio** at that time period had access to various forms of telecommunication devices, i.e. fixed and mobile phone line services. The level of income per capita for that same time period was equal to **$32,498.00**. The correlation coefficient measurement for these two variables is equal to 0.77669 (rounded to 5 decimal places), this means that 77.669% of the reason for which one variable moves in a specific direction can be explained by the variation in the other variable i.e. based upon the statistics provided 77.669% of the variation in the level of income per capita in the state of **Ohio** at that time period could be explained by the variation in the level of tele-accessibility, or access to telecommunications.

The state of **Indiana** had the fourteenth highest level of tele-accessibility for the year 2005, at a rate of 1.10973, a measurement which was rounded off to five decimal places. This means that on average 110.973% of the residents of **Indiana** at that time period had access to various forms of telecommunication devices, i.e. fixed and mobile phone line services. The level of income per capita for that same time period was equal to **$31,302.00**. The correlation coefficient measurement for these two variables is equal to 0.77669 (rounded to 5 decimal places), this means that 77.669% of the reason for which one variable moves in a specific direction can be explained by the variation in the other variable i.e. based upon the statistics provided 77.669% of the variation in the level of income per capita in the state of **Indiana** at that time period could be explained by the variation in the level of tele-accessibility, or access to telecommunications.

The state of **Utah** had the lowest level of tele-accessibility for the year 2005, at a rate of 0.98826, a measurement which was rounded off to five decimal places. This means that on average 98.826% of the residents of **Utah** at that time period had access to various forms of telecommunication devices, i.e. fixed and mobile phone line services. The level of income per capita for that same time period was equal to **$28,599.00**. The correlation coefficient measurement for these two variables is equal to 0.77669 (rounded to 5 decimal places), this means that 77.669% of the reason for which one variable moves in a specific direction can be explained by the variation in the other variable i.e. based upon the statistics provided 77.669% of the variation in the level of income per capita in the

state of **Utah** at that time period could be explained by the variation in the level of tele-accessibility, or access to telecommunications.

The level of Tele-Accessibility in the year 2006

The state of **District of Columbia** had the highest level of tele-accessibility for the year 2006, at a rate of 2.74503, a measurement which was rounded off to five decimal places. This means that on average 274.503% of the residents of **District of Columbia** at that time period had access to various forms of telecommunication devices, i.e. fixed and mobile phone line services. The level of income per capita for that same time period was equal to **$60,080.00**. The correlation coefficient measurement for these two variables is equal to 0.74439 (rounded to 5 decimal places), this means that 74.439 % of the reason for which one variable moves in a specific direction can be explained by the variation in the other variable i.e. based upon the statistics provided 74.439 % of the variation in the level of income per capita in the state of **District of Columbia** at that time period could be explained by the variation in the level of tele-accessibility, or access to telecommunications.

The state of **New Jersey** had the second highest level of tele-accessibility for the year 2006, at a rate of 1.39171, a measurement which was rounded off to five decimal places. This means that on average 139.171% of the residents of **New Jersey** at that time period had access to various forms of telecommunication devices, i.e. fixed and mobile phone line services. The level of income per capita for that same time period was equal to

$47,655.00. The correlation coefficient measurement for these two variables is equal to 0.74439 (rounded to 5 decimal places), this means that 74.439% of the reason for which one variable moves in a specific direction can be explained by the variation in the other variable i.e. based upon the statistics provided 74.439% of the variation in the level of income per capita in the state of **New Jersey** at that time period could be explained by the variation in the level of tele-accessibility, or access to telecommunications

The state of **California** had the third highest level of tele-accessibility for the year 2006, at a rate of 1.31007, a measurement which was rounded off to five decimal places. This means that on average 131.007% of the residents of **California** at that time period had access to various forms of telecommunication devices, i.e. fixed and mobile phone line services. The level of income per capita for that same time period was equal to **$41,404.00**. The correlation coefficient measurement for these two variables is equal to 0.74439 (rounded to 5 decimal places), this means that 74.439% of the reason for which one variable moves in a specific direction can be explained by the variation in the other variable i.e. based upon the statistics provided 74.439% of the variation in the level of income per capita in the state of **California** at that time period could be explained by the variation in the level of tele-accessibility, or access to telecommunications

The state of **Florida** had the fourth highest level of tele-accessibility for the year 2006, at a rate of 1.29316, a measurement which was rounded off to five decimal places. This means that on average 129.316% of the residents of **Florida** at that time period had access to various forms of telecommunication devices, i.e. fixed and mobile phone line services. The level of income per capita for that same time period was equal to **$38,308.00**. The correlation coefficient measurement for these two variables is equal to 0.74439 (rounded to 5 decimal places), this means that 74.439% of the reason for which one variable moves in a specific direction can be explained by the variation in the other variable i.e. based upon the statistics provided 74.439% of the variation in the level of income per capita in the state of **Florida** at that time period could be explained by the variation in the level of tele-accessibility, or access to telecommunications.

The state of **Connecticut** had the fifth highest level of tele-accessibility for the year 2006, at a rate of 1.27305, a measurement which was rounded off to five decimal places. This means that on average 127.305% of the residents of **Connecticut** at that time period had access to various forms of telecommunication devices, i.e. fixed and mobile phone line services. The level of income per capita for that same time period was equal to **$52,702.00**. The correlation coefficient measurement for these two variables is equal to 0.74439 (rounded to 5 decimal places), this means that 74.439% of the reason for which one variable moves in a specific direction can be explained by the variation in the other

variable i.e. based upon the statistics provided 74.439% of the variation in the level of income per capita in the state of **Connecticut** at that time period could be explained by the variation in the level of tele-accessibility, or access to telecommunications.

The state of **Massachusetts** had the sixth highest level of tele-accessibility for the year 2006, at a rate of 1.26015, a measurement which was rounded off to five decimal places. This means that on average 126.015% of the residents of **Massachusetts** at that time period had access to various forms of telecommunication devices, i.e. fixed and mobile phone line services. The level of income per capita for that same time period was equal to **$47,330.00**. The correlation coefficient measurement for these two variables is equal to 0.74439 (rounded to 5 decimal places), this means that 74.439% of the reason for which one variable moves in a specific direction can be explained by the variation in the other variable i.e. based upon the statistics provided 74.439% of the variation in the level of income per capita in the state of **Massachusetts** at that time period could be explained by the variation in the level of tele-accessibility, or access to telecommunications.

The state of **Tennessee** had the seventh highest level of tele-accessibility for the year 2006, at a rate of 1.24127, a measurement which was rounded off to five decimal places. This means that on average 124.127% of the residents of **Tennessee** at that time period had access to various forms of telecommunication devices, i.e. fixed and mobile phone line services. The level of income per capita for that same time period was equal to **$32,986.00**. The correlation coefficient measurement for these two variables is equal to 0.74439 (rounded to 5 decimal places), this means that 74.439% of the reason for which one variable moves in a specific direction can be explained by the variation in the other variable i.e. based upon the statistics provided 74.439% of the variation in the level of income per capita in the state of **Tennessee** at that time period could be explained by the variation in the level of tele-accessibility, or access to telecommunications.

The state of **Illinois** had the eighth highest level of tele-accessibility for the year 2006, at a rate of 1.22751, a measurement which was rounded off to five decimal places. This means that on average 122.751% of the residents of **Illinois** at that time period had access to various forms of telecommunication devices, i.e. fixed and mobile phone line services. The level of income per capita for that same time period was equal to **$39,549.00**. The correlation coefficient measurement for these two variables is equal to 0.74439 (rounded to 5 decimal places), this means that 74.439% of the reason for which one variable moves in a specific direction can be explained by the variation in the other variable i.e. based upon the statistics provided 74.439% of the variation in the level of income per capita in the state of **Illinois** at that time period could be explained by the variation in the level of tele-accessibility, or access to telecommunications.

The state of **New York** had the ninth highest level of tele-accessibility for the year 2006, at a rate of 1.22736, a measurement which was rounded off to five decimal places. This means that on average 122.736% of the residents of **New York** at that time period had access to various forms of telecommunication devices, i.e. fixed and mobile phone line services. The level of income per capita for that same time period was equal to **$43,973.00**. The correlation coefficient measurement for these two variables is equal to 0.74439 (rounded to 5 decimal places), this means that 74.439% of the reason for which one variable moves in a specific direction can be explained by the variation in the other variable i.e. based upon the statistics provided 74.439% of the variation in the level of income per capita in the state of **New York** at that time period could be explained by the variation in the level of tele-accessibility, or access to telecommunications.

The state of **Virginia** had the tenth highest level of tele-accessibility for the year 2006, at a rate of 1.20135, a measurement which was rounded off to five decimal places. This means that on average 120.135% of the residents of **Virginia** at that time period had access to various forms of telecommunication devices, i.e. fixed and mobile phone line services. The level of income per capita for that same time period was equal to **$41,367.00**. The correlation coefficient measurement for these two variables is equal to 0.74439 (rounded to 5 decimal places), this means that 74.439% of the reason for which one variable moves in a specific direction can be explained by the variation in the other variable i.e. based upon the statistics provided 74.439% of the variation in the level of income per capita in the state of **Virginia** at that time period could be explained by the variation in the level of tele-accessibility, or access to telecommunications.

The state of **Pennsylvania** had the eleventh highest level of tele-accessibility for the year 2006, at a rate of 1.19886, a measurement which was rounded off to five decimal places. This means that on average 119.886% of the residents of **Pennsylvania** at that time period had access to various forms of telecommunication devices, i.e. fixed and mobile phone line services. The level of income per capita for that same time period was equal to **$37,326.00**. The correlation coefficient measurement for these two variables is equal to 0.74439 (rounded to 5 decimal places), this means that 74.439% of the reason for which one variable moves in a specific direction can be explained by the variation in the other variable i.e. based upon the statistics provided 74.439% of the variation in the level of income per capita in the state of **Pennsylvania** at that time period could be explained by the variation in the level of tele-accessibility, or access to telecommunications.

The state of **Ohio** had the twelfth highest level of tele-accessibility for the year 2006, at a rate of 1.16364, a measurement which was rounded off to five decimal places. This means that on average 116.364% of the residents of **Ohio** at that time period had access to various forms of telecommunication devices, i.e. fixed and mobile phone line services. The level of income per capita for that same time period was equal to **$34,093.00**. The

correlation coefficient measurement for these two variables is equal to 0.74439 (rounded to 5 decimal places), this means that 74.439% of the reason for which one variable moves in a specific direction can be explained by the variation in the other variable i.e. based upon the statistics provided 74.439% of the variation in the level of income per capita in the state of **Ohio** at that time period could be explained by the variation in the level of tele-accessibility, or access to telecommunications.

The state of **Michigan** had the thirteenth highest level of tele-accessibility for the year 2006, at a rate of 1.14523, a measurement which was rounded off to five decimal places. This means that on average 114.523% of the residents of **Michigan** at that time period had access to various forms of telecommunication devices, i.e. fixed and mobile phone line services. The level of income per capita for that same time period was equal to **$33,198.00**. The correlation coefficient measurement for these two variables is equal to 0.74439 (rounded to 5 decimal places), this means that 74.439% of the reason for which one variable moves in a specific direction can be explained by the variation in the other variable i.e. based upon the statistics provided 74.439% of the variation in the level of income per capita in the state of **Michigan** at that time period could be explained by the variation in the level of tele-accessibility, or access to telecommunications.

The state of **Indiana** had the fourteenth highest level of tele-accessibility for the year 2006, at a rate of 1.11779, a measurement which was rounded off to five decimal places. This means that on average 111.779% of the residents of **Indiana** at that time period had access to various forms of telecommunication devices, i.e. fixed and mobile phone line services. The level of income per capita for that same time period was equal to **$32,881.00**. The correlation coefficient measurement for these two variables is equal to 0.74439 (rounded to 5 decimal places), this means that 74.439% of the reason for which one variable moves in a specific direction can be explained by the variation in the other variable i.e. based upon the statistics provided 74.439% of the variation in the level of income per capita in the state of **Indiana** at that time period could be explained by the variation in the level of tele-accessibility, or access to telecommunications.

The state of **Utah** had the lowest level of tele-accessibility for the year 2006, at a rate of 1.01681, a measurement which was rounded off to five decimal places. This means that on average 101.681% of the residents of **Utah** at that time period had access to various forms of telecommunication devices, i.e. fixed and mobile phone line services. The level of income per capita for that same time period was equal to **$30,320.00**. The correlation coefficient measurement for these two variables is equal to 0.74439 (rounded to 5 decimal places), this means that 74.439% of the reason for which one variable moves in a specific direction can be explained by the variation in the other variable i.e. based upon the statistics provided 74.439% of the variation in the level of income per capita in the

state of **Utah** at that time period could be explained by the variation in the level of tele-accessibility, or access to telecommunications.

The level of Tele-Accessibility in the year 2007

The state of **District of Columbia** had the highest level of tele-accessibility for the year 2007, at a rate of 3.30011, a measurement which was rounded off to five decimal places. This means that on average 330.011% of the residents of **District of Columbia** at that time period had access to various forms of telecommunication devices, i.e. fixed and mobile phone line services. The level of income per capita for that same time period was equal to **$63,881.00**. The correlation coefficient measurement for these two variables is equal to 0.72998 (rounded to 5 decimal places), this means that 72.998% of the reason for which one variable moves in a specific direction can be explained by the variation in the other variable i.e. based upon the statistics provided 72.998% of the variation in the level of income per capita in the state of **District of Columbia** at that time period could be explained by the variation in the level of tele-accessibility, or access to telecommunications.

The state of **New Jersey** had the second highest level of tele-accessibility for the year 2007, at a rate of 1.46611, a measurement which was rounded off to five decimal places. This means that on average 146.611% of the residents of **New Jersey** at that time period had access to various forms of telecommunication devices, i.e. fixed and mobile phone line services. The level of income per capita for that same time period was equal to

$50,265.00. The correlation coefficient measurement for these two variables is equal to 0.72998 (rounded to 5 decimal places), this means that 72.998% of the reason for which one variable moves in a specific direction can be explained by the variation in the other variable i.e. based upon the statistics provided 72.998% of the variation in the level of income per capita in the state of **New Jersey** at that time period could be explained by the variation in the level of tele-accessibility, or access to telecommunications

The state of **California** had the third highest level of tele-accessibility for the year 2007, at a rate of 1.42405, a measurement which was rounded off to five decimal places. This means that on average 142.405% of the residents of **California** at that time period had access to various forms of telecommunication devices, i.e. fixed and mobile phone line services. The level of income per capita for that same time period was equal to **$43,221.00**. The correlation coefficient measurement for these two variables is equal to 0.72998 (rounded to 5 decimal places), this means that 72.998% of the reason for which one variable moves in a specific direction can be explained by the variation in the other variable i.e. based upon the statistics provided 72.998% of the variation in the level of income per capita in the state of **California** at that time period could be explained by the variation in the level of tele-accessibility, or access to telecommunications

The state of **Virginia** had the fourth highest level of tele-accessibility for the year 2007, at a rate of 1.40403, a measurement which was rounded off to five decimal places. This means that on average 140.403% of the residents of **Virginia** at that time period had access to various forms of telecommunication devices, i.e. fixed and mobile phone line services. The level of income per capita for that same time period was equal to **$43,275.00**. The correlation coefficient measurement for these two variables is equal to 0.72998 (rounded to 5 decimal places), this means that 72.998% of the reason for which one variable moves in a specific direction can be explained by the variation in the other variable i.e. based upon the statistics provided 72.998% of the variation in the level of income per capita in the state of **Virginia** at that time period could be explained by the variation in the level of tele-accessibility, or access to telecommunications.

The state of **Connecticut** had the fifth highest level of tele-accessibility for the year 2007, at a rate of 1.38517, a measurement which was rounded off to five decimal places. This means that on average 138.517% of the residents of **Connecticut** at that time period had access to various forms of telecommunication devices, i.e. fixed and mobile phone line services. The level of income per capita for that same time period was equal to **$55,609.00**. The correlation coefficient measurement for these two variables is equal to 0.72998 (rounded to 5 decimal places), this means that 72.998% of the reason for which

one variable moves in a specific direction can be explained by the variation in the other variable i.e. based upon the statistics provided 72.998% of the variation in the level of income per capita in the state of **Connecticut** at that time period could be explained by the variation in the level of tele-accessibility, or access to telecommunications.

The state of **Massachusetts** had the sixth highest level of tele-accessibility for the year 2007, at a rate of 1.38242, a measurement which was rounded off to five decimal places. This means that on average 138.242% of the residents of **Massachusetts** at that time period had access to various forms of telecommunication devices, i.e. fixed and mobile phone line services. The level of income per capita for that same time period was equal to **$49,885.00**. The correlation coefficient measurement for these two variables is equal to 0.72998 (rounded to 5 decimal places), this means that 72.998% of the reason for which one variable moves in a specific direction can be explained by the variation in the other variable i.e. based upon the statistics provided 72.998% of the variation in the level of income per capita in the state of **Massachusetts** at that time period could be explained by the variation in the level of tele-accessibility, or access to telecommunications.

The state of **Florida** had the seventh highest level of tele-accessibility for the year 2007, at a rate of 1.38196, a measurement which was rounded off to five decimal places. This means that on average 138.196% of the residents of **Florida** at that time period had access to various forms of telecommunication devices, i.e. fixed and mobile phone line services. The level of income per capita for that same time period was equal to **$39,204.00**. The correlation coefficient measurement for these two variables is equal to 0.72998 (rounded to 5 decimal places), this means that 72.998% of the reason for which one variable moves in a specific direction can be explained by the variation in the other variable i.e. based upon the statistics provided 72.998% of the variation in the level of income per capita in the state of **Florida** at that time period could be explained by the variation in the level of tele-accessibility, or access to telecommunications.

The state of **New York** had the eighth highest level of tele-accessibility for the year 2007, at a rate of 1.34749, a measurement which was rounded off to five decimal places. This means that on average 134.749% of the residents of **New York** at that time period had access to various forms of telecommunication devices, i.e. fixed and mobile phone line services. The level of income per capita for that same time period was equal to **$47,612.00**. The correlation coefficient measurement for these two variables is equal to 0.72998 (rounded to 5 decimal places), this means that 72.998% of the reason for which one variable moves in a specific direction can be explained by the variation in the other variable i.e. based upon the statistics provided 72.998% of the variation in the level of income per capita in the state of **New York** at that time period could be explained by the variation in the level of tele-accessibility, or access to telecommunications.

The state of **Pennsylvania** had the ninth highest level of tele-accessibility for the year 2007, at a rate of 1.33157, a measurement which was rounded off to five decimal places. This means that on average 133.157% of the residents of **Pennsylvania** at that time period had access to various forms of telecommunication devices, i.e. fixed and mobile phone line services. The level of income per capita for that same time period was equal to **$39,058.00**. The correlation coefficient measurement for these two variables is equal to 0.72998 (rounded to 5 decimal places), this means that 72.998% of the reason for which one variable moves in a specific direction can be explained by the variation in the other variable i.e. based upon the statistics provided 72.998% of the variation in the level of income per capita in the state of **Pennsylvania** at that time period could be explained by the variation in the level of tele-accessibility, or access to telecommunications.

The state of **Illinois** had the tenth highest level of tele-accessibility for the year 2007, at a rate of 1.32044, a measurement which was rounded off to five decimal places. This means that on average 132.044% of the residents of **Illinois** at that time period had access to various forms of telecommunication devices, i.e. fixed and mobile phone line services. The level of income per capita for that same time period was equal to **$41,569.00**. The correlation coefficient measurement for these two variables is equal to 0.72998 (rounded to 5 decimal places), this means that 72.998% of the reason for which one variable moves in a specific direction can be explained by the variation in the other variable i.e. based upon the statistics provided 72.998% of the variation in the level of income per capita in the state of **Illinois** at that time period could be explained by the variation in the level of tele-accessibility, or access to telecommunications.

The state of **Tennessee** had the eleventh highest level of tele-accessibility for the year 2007, at a rate of 1.30768, a measurement which was rounded off to five decimal places. This means that on average 130.768% of the residents of **Tennessee** at that time period had access to various forms of telecommunication devices, i.e. fixed and mobile phone line services. The level of income per capita for that same time period was equal to **$34,287.00**. The correlation coefficient measurement for these two variables is equal to 0.72998 (rounded to 5 decimal places), this means that 72.998% of the reason for which one variable moves in a specific direction can be explained by the variation in the other variable i.e. based upon the statistics provided 72.998% of the variation in the level of income per capita in the state of **Tennessee** at that time period could be explained by the variation in the level of tele-accessibility, or access to telecommunications.

The state of **Ohio** had the twelfth highest level of tele-accessibility for the year 2007, at a rate of 1.28155, a measurement which was rounded off to five decimal places. This means that on average 128.155% of the residents of **Ohio** at that time period had access to various forms of telecommunication devices, i.e. fixed and mobile phone line services. The level of income per capita for that same time period was equal to **$35,307.00**. The

correlation coefficient measurement for these two variables is equal to 0.72998 (rounded to 5 decimal places), this means that 72.998% of the reason for which one variable moves in a specific direction can be explained by the variation in the other variable i.e. based upon the statistics provided 72.998% of the variation in the level of income per capita in the state of **Ohio** at that time period could be explained by the variation in the level of tele-accessibility, or access to telecommunications.

The state of **Michigan** had the thirteenth highest level of tele-accessibility for the year 2007, at a rate of 1.23119, a measurement which was rounded off to five decimal places. This means that on average 123.119% of the residents of **Michigan** at that time period had access to various forms of telecommunication devices, i.e. fixed and mobile phone line services. The level of income per capita for that same time period was equal to **$34,188.00**. The correlation coefficient measurement for these two variables is equal to 0.72998 (rounded to 5 decimal places), this means that 72.998% of the reason for which one variable moves in a specific direction can be explained by the variation in the other variable i.e. based upon the statistics provided 72.998% of the variation in the level of income per capita in the state of **Michigan** at that time period could be explained by the variation in the level of tele-accessibility, or access to telecommunications.

The state of **Indiana** had the fourteenth highest level of tele-accessibility for the year 2007, at a rate of 1.20002, a measurement which was rounded off to five decimal places. This means that on average 120.002% of the residents of **Indiana** at that time period had access to various forms of telecommunication devices, i.e. fixed and mobile phone line services. The level of income per capita for that same time period was equal to **$33,756.00**. The correlation coefficient measurement for these two variables is equal to 0.72998 (rounded to 5 decimal places), this means that 72.998% of the reason for which one variable moves in a specific direction can be explained by the variation in the other variable i.e. based upon the statistics provided 72.998% of the variation in the level of income per capita in the state of **Indiana** at that time period could be explained by the variation in the level of tele-accessibility, or access to telecommunications.

The state of **Utah** had the lowest level of tele-accessibility for the year 2007, at a rate of 1.11887, a measurement which was rounded off to five decimal places. This means that on average 111.887% of the residents of **Utah** at that time period had access to various forms of telecommunication devices, i.e. fixed and mobile phone line services. The level of income per capita for that same time period was equal to **$31,739.00**. The correlation coefficient measurement for these two variables is equal to 0.72998 (rounded to 5 decimal places), this means that 72.998% of the reason for which one variable moves in a specific direction can be explained by the variation in the other variable i.e. based upon the statistics provided 72.998% of the variation in the level of income per capita in the

state of **Utah** at that time period could be explained by the variation in the level of tele-accessibility, or access to telecommunications.

The level of Tele-Accessibility in the State of California for the years 2000-2007

For the state of **California** if one looks at the bar charts which represent the changes in the rate of tele-accessibility and income per capita over the time period of 2000-2007, you are able to see that both variables move in the same direction, and thus are positively correlated. The correlation coefficient measurement that exists between these two variables for the same time period is equal to a rate of 0.97167 a measurement which was rounded off to five decimal places, or states that 97.167 % of the variation in the level of income per capita in the state of **California** may be explained by its access to telecommunications.

The level of Tele-Accessibility in the State of Connecticut for the years 2000-2007

For the state of **Connecticut** if one looks at the bar charts which represent the changes in the rate of tele-accessibility and income per capita over the time period of 2000-2007, you are able to see that both variables move in the same direction, and thus are positively correlated. The correlation coefficient measurement that exists between these two variables for the same time period is equal to a rate of 0.90445 a measurement which was rounded off to five decimal places, or states that 90.445% of the variation in the level of income per capita in the state of **Connecticut** may be explained by its access to telecommunications.

The level of Tele-Accessibility in the District of Columbia for the years 2000-2007

For the state of **District of Columbia** if one looks at the bar charts which represent the changes in the rate of tele-accessibility and income per capita over the time period of 2000-2007, you are able to see that both variables move in the same direction, and thus are positively correlated. The correlation coefficient measurement that exists between these two variables for the same time period is equal to a rate of 0.87001 a measurement which was rounded off to five decimal places, or states that 87.001% of the variation in the level of income per capita in the state of **District of Columbia** may be explained by its access to telecommunications.

The level of Tele-Accessibility in the State of Florida for the years 2000-2007

For the state of **Florida** if one looks at the bar charts which represent the changes in the rate of tele-accessibility and income per capita over the time period of 2000-2007, you are able to see that both variables move in the same direction, and thus are positively correlated. The correlation coefficient measurement that exists between these two variables for the same time period is equal to a rate of 0.81745 a measurement which was rounded off to five decimal places, or states that 81.745% of the variation in the level of income per capita in the state of **Florida** may be explained by its access to telecommunications.

The level of Tele-Accessibility in the State of Illinois for the years 2000-2007

For the state of **Illinois** if one looks at the bar charts which represent the changes in the rate of tele-accessibility and income per capita over the time period of 2000-2007, you are able to see that both variables move in the same direction, and thus are positively correlated. The correlation coefficient measurement that exists between these two variables for the same time period is equal to a rate of 0.90901 a measurement which was rounded off to five decimal places, or states that 90.901% of the variation in the level of income per capita in the state of **Illinois** may be explained by its access to telecommunications.

The level of Tele-Accessibility in the State of Indiana for the years 2000-2007

For the state of **Indiana** if one looks at the bar charts which represent the changes in the rate of tele-accessibility and income per capita over the time period of 2000-2007, you are able to see that both variables move in the same direction, and thus are positively correlated. The correlation coefficient measurement that exists between these two variables for the same time period is equal to a rate of 0.98130 a measurement which was rounded off to five decimal places, or states that 98.130% of the variation in the level of income per capita in the state of **Indiana** may be explained by its access to telecommunications.

The level of Tele-Accessibility in the State of Massachusetts for the years 2000-2007

For the state of **Massachusetts** if one looks at the bar charts which represent the changes in the rate of tele-accessibility and income per capita over the time period of 2000-2007, you are able to see that both variables move in the same direction, and thus are positively correlated. The correlation coefficient measurement that exists between these two variables for the same time period is equal to a rate of 0.85409 a measurement which was rounded off to five decimal places, or states that 85.409% of the variation in the level of income per capita in the state of **Massachusetts** may be explained by its access to telecommunications.

The level of Tele-Accessibility in the State of Michigan for the years 2000-2007

For the state of **Michigan** if one looks at the bar charts which represent the changes in the rate of tele-accessibility and income per capita over the time period of 2000-2007, you are able to see that both variables move in the same direction, and thus are positively correlated. The correlation coefficient measurement that exists between these two variables for the same time period is equal to a rate of 0.90474 a measurement which was rounded off to five decimal places, or states that 90.474% of the variation in the level of income per capita in the state of **Michigan** may be explained by its access to telecommunications.

The level of Tele-Accessibility in the State of New Jersey for the years 2000-2007

For the state of **New Jersey** if one looks at the bar charts which represent the changes in the rate of tele-accessibility and income per capita over the time period of 2000-2007, you are able to see that both variables move in the same direction, and thus are positively correlated. The correlation coefficient measurement that exists between these two variables for the same time period is equal to a rate of 0.75574 a measurement which was rounded off to five decimal places, or states that 75.574% of the variation in the level of income per capita in the state of **New Jersey** may be explained by its access to telecommunications.

The level of Tele-Accessibility in the State of New York for the years 2000-2007

For the state of **New York** if one looks at the bar charts which represent the changes in the rate of tele-accessibility and income per capita over the time period of 2000-2007, you are able to see that both variables move in the same direction, and thus are positively correlated. The correlation coefficient measurement that exists between these two variables for the same time period is equal to a rate of 0.93578 a measurement which was rounded off to five decimal places, or states that 93.578% of the variation in the level of income per capita in the state of **New York** may be explained by its access to telecommunications.

The level of Tele-Accessibility in the State of Pennsylvania for the years 2000-2007

For the state of **Pennsylvania** if one looks at the bar charts which represent the changes in the rate of tele-accessibility and income per capita over the time period of 2000-2007, you are able to see that both variables move in the same direction, and thus are positively correlated. The correlation coefficient measurement that exists between these two variables for the same time period is equal to a rate of 0.95083 a measurement which was rounded off to five decimal places, or states that 95.083% of the variation in the level of income per capita in the state of **Pennsylvania** may be explained by its access to telecommunications.

The level of Tele-Accessibility in the State of Tennessee for the years 2000-2007

For the state of **Tennessee** if one looks at the bar charts which represent the changes in the rate of tele-accessibility and income per capita over the time period of 2000-2007, you are able to see that both variables move in the same direction, and thus are positively correlated. The correlation coefficient measurement that exists between these two variables for the same time period is equal to a rate of 0.99182 a measurement which was rounded off to five decimal places, or states that 99.182% of the variation in the level of income per capita in the state of **Tennessee** may be explained by its access to telecommunications.

The level of Tele-Accessibility in the State of Virginia for the years 2000-2007

For the state of **Virginia** if one looks at the bar charts which represent the changes in the rate of tele-accessibility and income per capita over the time period of 2000-2007, you are able to see that both variables move in the same direction, and thus are positively correlated. The correlation coefficient measurement that exists between these two variables for the same time period is equal to a rate of 0.86839 a measurement which was rounded off to five decimal places, or states that 86.839% of the variation in the level of income per capita in the state of **Virginia** may be explained by its access to telecommunications.

The level of Tele-Accessibility in the State of Utah for the years 2000-2007

For the state of **Utah** if one looks at the bar charts which represent the changes in the rate of tele-accessibility and income per capita over the time period of 2000-2007, you are able to see that both variables move in the same direction, and thus are positively correlated. The correlation coefficient measurement that exists between these two variables for the same time period is equal to a rate of 0.89677 a measurement which was rounded off to five decimal places, or states that 89.677% of the variation in the level of income per capita in the state of **Utah** may be explained by its access to telecommunications.

The level of Tele-Accessibility in the State of Ohio for the years 2000-2007

For the state of **Ohio** if one looks at the bar charts which represent the changes in the rate of tele-accessibility and income per capita over the time period of 2000-2007, you are able to see that both variables move in the same direction, and thus are positively correlated. The correlation coefficient measurement that exists between these two variables for the same time period is equal to a rate of 0.96291 a measurement which was rounded off to five decimal places, or states that 96.291% of the variation in the level of income per capita in the state of **Ohio** may be explained by its access to telecommunications.

Conclusion

For the states which are represented in this study one is able to observe by taking a look at the bar charts which represent the changes in the rate of tele-accessibility and income per capita over the time period of 2000-2007, that both variables move in the same direction, and thus are positively correlated. The correlation that exists between these two variables for the same time period range from a rate of 0.47636 to 0.77669, a measurement which was rounded off to five decimal places, or states that between 47.636% and 77.669% of the variation in the level of income per capita in the various states may be explained by its access to telecommunications.

One observation which I have noticed is that once the level of tele-accessibility reaches a rate of 1.0 or higher, the strength of the correlation factor between the two variables are diminished or reduced. This means that there are other variables which play a larger contributing role to the continued growth in income per capita than the tele-accessibility measurement, beyond this point.

<u>Appendix</u>

1. The excel spreadsheets and bar charts which show the numbers from which the conclusions were derived.

	State	Population	Number of Fixed Lines	Number of Mobile Lines	Tele - Accessibility (Fixed + Mobile lines /Population)	Income per Capita		Income per Capita / 10,000	
					Year - 2000				
1	District of Columbia	571,744	987,412	333,815	2.310871649	$	37,383.00	$	3.74
2	New Jersey	8,430,921	7,000,131	2,750,024	1.156475669	$	36,983.00	$	3.70
3	California	33,994,571	24,754,207	12,283,369	1.08951444	$	32,275.00	$	3.23
4	Massachusetts	6,363,015	4,698,536	2,228,169	1.088588507	$	37,992.00	$	3.80
5	Connecticut	3,411,726	2,574,205	1,136,618	1.087667357	$	40,640.00	$	4.06
6	Florida	16,047,118	12,104,421	4,983,478	1.064857814	$	28,145.00	$	2.81
7	Illinois	12,437,645	8,740,081	4,309,660	1.049213175	$	32,259.00	$	3.23
8	Pennsylvania	12,285,504	8,871,784	3,850,372	1.03554205	$	29,539.00	$	2.95
9	Michigan	9,955,308	6,722,255	3,423,535	1.019133712	$	29,612.00	$	2.96
10	New York	18,998,044	13,689,883	5,016,524	0.984649104	$	34,547.00	$	3.45
11	Virginia	7,104,533	4,469,865	2,447,687	0.973681451	$	31,162.00	$	3.12
12	Tennessee	5,703,243	3,525,455	1,876,444	0.947162693	$	26,239.00	$	2.62
13	Ohio	11,363,844	7,211,041	3,278,960	0.923103221	$	28,400.00	$	2.84
14	Indiana	6,091,649	3,753,645	1,717,378	0.898118555	$	27,011.00	$	2.70
15	Utah	2,244,314	1,286,615	692,006	0.88161505	$	23,907.00	$	2.39
	Total	155,003,179							
	Average		7,359,302	3,355,203	1.10068	$	31,739.60	$	3.17
	Correlation Coefficient -	0.47635533							

Websites used:

http://www.census.gov/popest/states/NST-ann-est.html
http://www.fcc.gov/Bureaus/Common_Carrier/Reports/FCC-State_Link/IAD/trend801.pdf
http://www.bea.gov/scb/pdf/2010/02%20February/DPages/0210dpg_i.pdf
http://www.bea.gov/newsreleases/relsarchivespi.htm
http://www.fcc.gov/Bureaus/Common_Carrier/Reports/FCC-State_Link/IAD/trend803.pdf
http://www.fcc.gov/Bureaus/Common_Carrier/Reports/FCC-State_Link/IAD/trend200.pdf

Year - 2000		
State	Tele - Accessibility (Fixed + Mobile lines /Population)	Income per Capita /10000
District of Columbia	2.310871649	$ 3.74
New Jersey	1.156475669	$ 3.70
California	1.08951444	$ 3.23
Massachusetts	1.088588507	$ 3.80
Connecticut	1.087667357	$ 4.06
Florida	1.064857814	$ 2.81
Illinois	1.049213175	$ 3.23
Pennsylvania	1.03554205	$ 2.95
Michigan	1.019133712	$ 2.96
New York	0.984649104	$ 3.45
Virginia	0.973681451	$ 3.12
Tennessee	0.947162693	$ 2.62
Ohio	0.923103221	$ 2.84
Indiana	0.898118555	$ 2.70
Utah	0.88161505	$ 2.39
Average	1.10068	
Correlation Coefficient -	0.476355325	

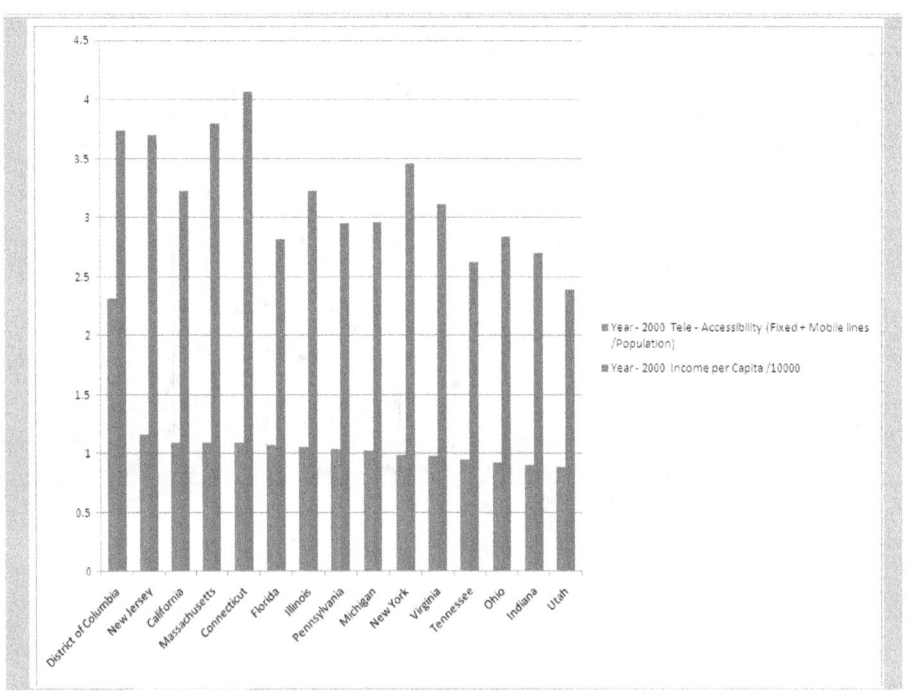

Legend:
- Year - 2000 Tele - Accessibility (Fixed + Mobile lines /Population)
- Year - 2000 Income per Capita /10000

	State	Population	Number of Fixed Lines	Number of Mobile Lines	Tele - Accessibility (Fixed + Mobile lines /Population)	Income per Capita	Income per Capita / 10,000
					Year - 2001		
1	District of Columbia	578,042	919,587	382,457	2.252507603	$ 40,150.00	$ 4.02
2	New Jersey	8,489,469	6,923,410	3,896,778	1.274542377	$ 38,509.00	$ 3.85
3	Florida	16,353,869	11,317,933	7,536,670	1.152913907	$ 28,947.00	$ 2.89
4	Massachusetts	6,411,730	4,410,394	2,753,685	1.11733947	$ 38,907.00	$ 3.89
5	Connecticut	3,428,433	2,406,704	1,418,367	1.115690754	$ 42,435.00	$ 4.24
6	Illinois	12,507,833	8,012,870	5,621,044	1.090030064	$ 33,023.00	$ 3.30
7	California	34,485,623	23,385,691	14,184,625	1.089448667	$ 32,702.00	$ 3.27
8	Virginia	7,191,304	4,760,302	3,059,420	1.08738582	$ 32,431.00	$ 3.24
9	New York	19,088,978	13,076,558	6,749,096	1.038591694	$ 36,019.00	$ 3.60
10	Pennsylvania	12,299,533	8,301,408	4,378,216	1.030902881	$ 30,720.00	$ 3.07
11	Michigan	10,006,093	6,149,365	4,071,091	1.021423247	$ 29,788.00	$ 2.98
12	Ohio	11,396,874	7,053,650	4,255,934	0.992340882	$ 28,816.00	$ 2.88
13	Tennessee	5,755,443	3,385,953	2,251,208	0.979448671	$ 26,988.00	$ 2.70
14	Indiana	6,124,967	3,803,634	1,781,247	0.911822219	$ 27,783.00	$ 2.78
15	Utah	2,291,250	1,172,443	833,492	0.875476268	$ 24,180.00	$ 2.42
	Total	156,409,441					
	Average		7,005,327	4,211,555	1.135324	$ 32,759.87	$ 3.28

Correlation Coefficient - 0.56316782

Websites used:

http://www.census.gov/popest/states/NST-ann-est.html
http://www.fcc.gov/Bureaus/Common_Carrier/Reports/FCC-State_Link/IAD/trend801.pdf
http://www.bea.gov/newsreleases/relsarchivespi.htm
http://hraunfoss.fcc.gov/edocs_public/attachmatch/DOC-284932A1.pdf
http://www.fcc.gov/Bureaus/Common_Carrier/Reports/FCC-State_Link/IAD/trend803.pdf

Year - 2001		
State	Tele - Accessibility (Fixed + Mobile lines /Population)	Income per Capita /10000
District of Columbia	2.252507603	$ 4.02
New Jersey	1.274542377	$ 3.85
Florida	1.152913907	$ 2.89
Massachusetts	1.11733947	$ 3.89
Connecticut	1.115690754	$ 4.24
Illinois	1.090030064	$ 3.30
California	1.089448667	$ 3.27
Virginia	1.08738582	$ 3.24
New York	1.038591694	$ 3.60
Pennsylvania	1.030902881	$ 3.07
Michigan	1.021423247	$ 2.98
Ohio	0.992340882	$ 2.88
Tennessee	0.979448671	$ 2.70
Indiana	0.911822219	$ 2.78
Utah	0.875476268	$ 2.42
Average	1.135324	
Correlation Coefficient -	0.563167817	

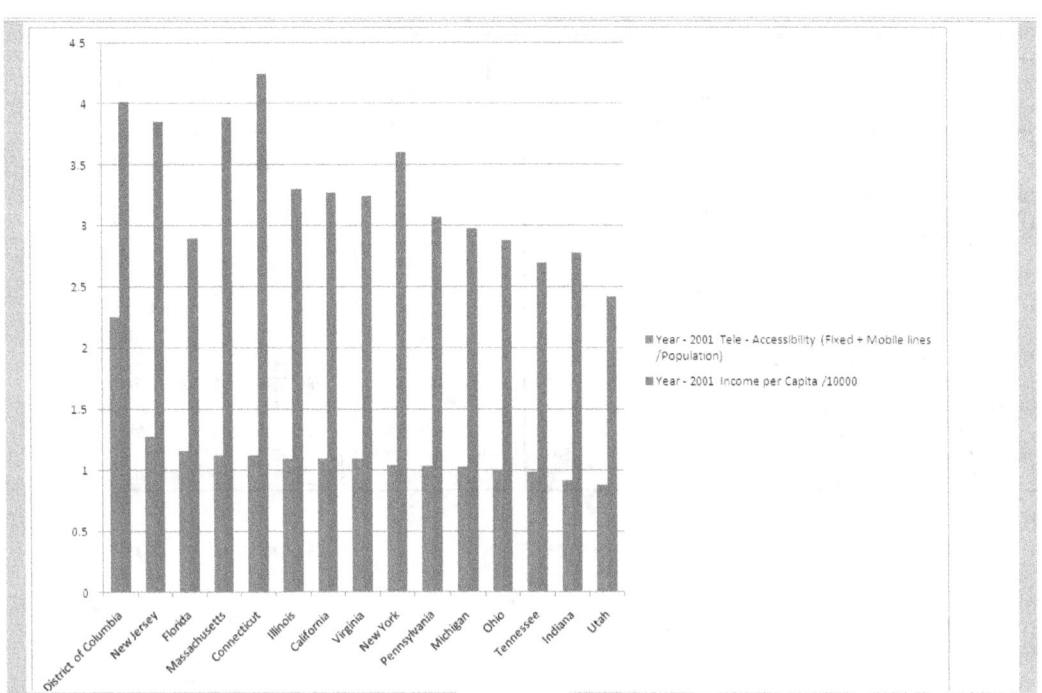

Legend:
- Year - 2001 Tele - Accessibility (Fixed + Mobile lines /Population)
- Year - 2001 Income per Capita /10000

					Tele - Accessibility		Income per
	State	Population	Number of Fixed Lines	Number of Mobile Lines	(Fixed + Mobile lines /Population)	Income per Capita		Capita / 10,000
1	District of Columbia	579,585	992,094	415,399	2.428449667	$	48,342.00	$ 4.83
2	New Jersey	8,544,115	6,580,282	4,531,457	1.300513745	$	40,427.00	$ 4.04
3	Florida	16,680,309	11,901,261	8,607,715	1.229532139	$	30,446.00	$ 3.04
4	Massachusetts	6,440,978	4,501,471	3,289,934	1.209661794	$	39,815.00	$ 3.98
5	Connecticut	3,448,382	2,499,908	1,577,873	1.182520092	$	43,173.00	$ 4.32
6	California	34,876,194	24,174,586	16,007,376	1.152131508	$	33,749.00	$ 3.37
7	Virginia	7,283,541	4,902,153	3,429,450	1.14389457	$	33,671.00	$ 3.37
8	Michigan	10,038,767	6,536,688	4,758,538	1.12516069	$	30,439.00	$ 3.04
9	Illinois	12,558,229	8,596,609	5,409,370	1.115282975	$	33,690.00	$ 3.37
10	Pennsylvania	12,326,302	8,573,098	4,987,067	1.100100014	$	31,998.00	$ 3.20
11	New York	19,161,873	12,821,422	7,915,526	1.082198384	$	36,574.00	$ 3.66
12	Tennessee	5,803,306	3,474,219	2,660,068	1.057033181	$	28,455.00	$ 2.85
13	Ohio	11,420,981	7,057,674	4,887,376	1.045886514	$	29,944.00	$ 2.99
14	Utah	2,334,473	1,269,413	970,854	0.95964571	$	24,977.00	$ 2.50
15	Indiana	6,149,007	3,744,405	2,032,290	0.939451687	$	28,783.00	$ 2.88
	Total	157,646,042						
	Average		7,175,019	4,765,353	1.20476	$	34,298.87	$ 3.43

Year - 2002

Correlation Coefficient - 0.74782585

Websites used:

http://www.census.gov/popest/states/NST-ann-est.html
http://www.fcc.gov/Bureaus/Common_Carrier/Reports/FCC-State_Link/IAD/trend801.pdf
http://www.bea.gov/newsreleases/relsarchivespi.htm
http://hraunfoss.fcc.gov/edocs_public/attachmatch/DOC-284932A1.pdf
http://www.fcc.gov/Bureaus/Common_Carrier/Reports/FCC-State_Link/IAD/trend803.pdf

Year - 2002		
State	Tele - Accessibility (Fixed + Mobile lines /Population)	Income per Capita /10000
District of Columbia	2.428449667	$ 4.83
New Jersey	1.300513745	$ 4.04
Florida	1.229532139	$ 3.04
Massachusetts	1.209661794	$ 3.98
Connecticut	1.182520092	$ 4.32
California	1.152131508	$ 3.37
Virginia	1.14389457	$ 3.37
Michigan	1.12516069	$ 3.04
Illinois	1.115282975	$ 3.37
Pennsylvania	1.100100014	$ 3.20
New York	1.082198384	$ 3.66
Tennessee	1.057033181	$ 2.85
Ohio	1.045886514	$ 2.99
Utah	0.95964571	$ 2.50
Indiana	0.939451687	$ 2.88
Average	1.20476	
Correlation Coefficient -	0.747825854	

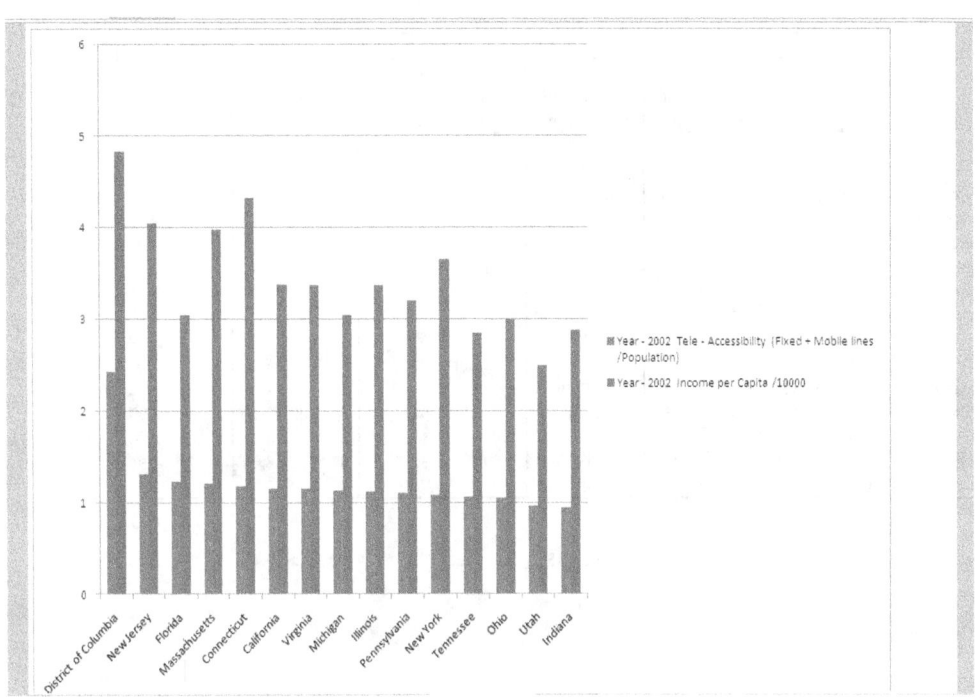

Legend:
- Year - 2002 Tele - Accessiblity (Fixed + Mobile lines /Population)
- Year - 2002 Income per Capita /10000

	State	Population	Number of Fixed Lines	Number of Mobile Lines	Tele - Accessibility (Fixed + Mobile lines /Population)	Income per Capita	Income per Capita / 10,000
				Year - 2003			
1	District of Columbia	577,777	947,171	520,182	2.539652842	$ 47,529.00	$ 4.75
2	New Jersey	8,583,481	6,399,743	5,392,240	1.373799627	$ 40,504.00	$ 4.05
3	Florida	16,981,183	11,671,497	10,252,348	1.291067	$ 31,364.00	$ 3.14
4	Massachusetts	6,451,637	4,407,964	3,506,039	1.226665883	$ 40,161.00	$ 4.02
5	Connecticut	3,467,673	2,449,918	1,791,944	1.223258941	$ 43,730.00	$ 4.37
6	California	35,251,107	23,692,322	18,892,619	1.208045495	$ 34,922.00	$ 3.49
7	Illinois	12,597,981	8,357,937	6,834,217	1.205919742	$ 34,569.00	$ 3.46
8	Virginia	7,373,694	4,759,521	3,879,582	1.171611271	$ 35,029.00	$ 3.50
9	Pennsylvania	12,357,524	8,261,544	5,681,653	1.128316401	$ 32,427.00	$ 3.24
10	New York	19,231,101	12,498,312	8,829,070	1.109004731	$ 36,165.00	$ 3.62
11	Michigan	10,066,351	6,204,267	4,889,269	1.102041445	$ 31,214.00	$ 3.12
12	Ohio	11,445,180	6,885,788	5,659,459	1.096116182	$ 30,698.00	$ 3.07
13	Tennessee	5,856,522	3,388,799	2,800,735	1.056861735	$ 29,026.00	$ 2.90
14	Indiana	6,181,789	3,675,394	2,456,509	0.991930168	$ 29,588.00	$ 2.96
15	Utah	2,379,938	1,254,259	1,094,563	0.98692571	$ 25,830.00	$ 2.58
	Total	158,802,938					
	Average		6,990,296	5,498,695	1.24741	$ 34,850.40	$ 3.49
	Correlation Coefficient -	0.73257499					

Websites used:

http://www.census.gov/popest/states/NST_ann-est.html
http://www.fcc.gov/Bureaus/Common_Carrier/Reports/FCC-State_Link/IAD/trend801.pdf
http://www.bea.gov/newsreleases/relsarchivespi.htm
http://hraunfoss.fcc.gov/edocs_public/attachmatch/DOC-284932A1.pdf
http://www.fcc.gov/Bureaus/Common_Carrier/Reports/FCC-State_Link/IAD/trend504.pdf

	Year - 2003	
State	Tele - Accessibility (Fixed + Mobile lines /Population)	Income per Capita /10000
District of Columbia	2.539652842	$ 4.75
New Jersey	1.373799627	$ 4.05
Florida	1.291067	$ 3.14
Massachusetts	1.226665883	$ 4.02
Connecticut	1.223258941	$ 4.37
California	1.208045495	$ 3.49
Illinois	1.205919742	$ 3.46
Virginia	1.171611271	$ 3.50
Pennsylvania	1.128316401	$ 3.24
New York	1.109004731	$ 3.62
Michigan	1.102041445	$ 3.12
Ohio	1.096116182	$ 3.07
Tennessee	1.056861735	$ 2.90
Indiana	0.991930168	$ 2.96
Utah	0.98692571	$ 2.58
Average	1.24741	
Correlation Coefficient -	0.732574986	

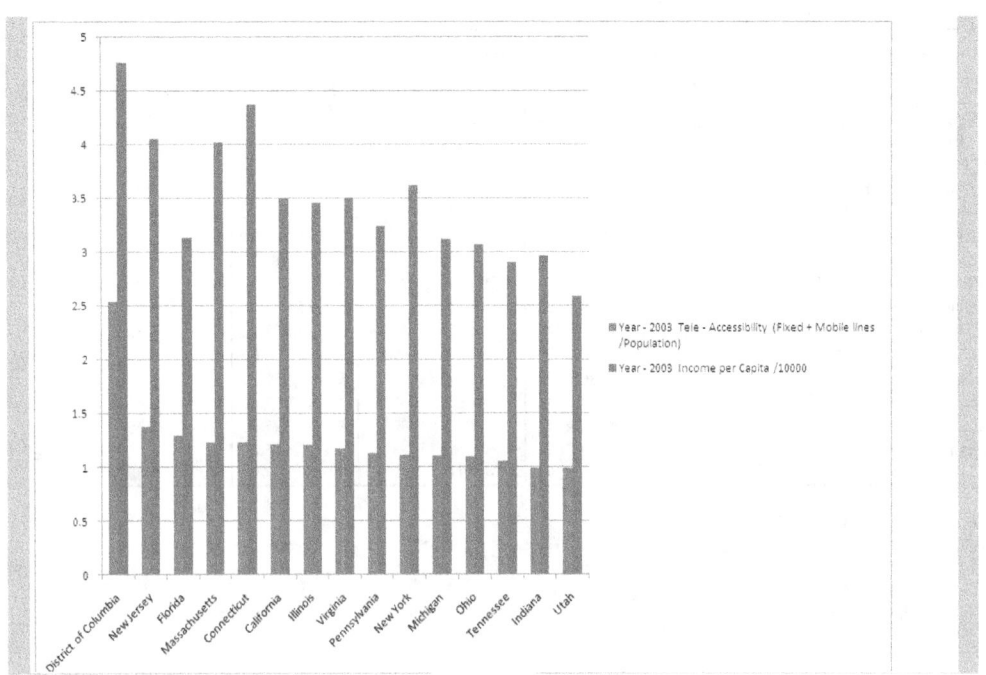

Legend:
- Year - 2003 Tele - Accessibility (Fixed + Mobile lines /Population)
- Year - 2003 Income per Capita /10000

	State	Population	Number of Fixed Lines	Number of Mobile Lines	Tele - Accessibility (Fixed + Mobile lines /Population)	Income per Capita	Income per Capita / 10,000
					Year - 2004		
1	District of Columbia	579,796	1,131,004	555,958	2.909578541	$ 51,458.00	$ 5.15
2	New Jersey	8,611,530	6,468,140	6,326,459	1.485752125	$ 42,406.00	$ 4.24
3	Florida	17,375,259	11,418,566	11,916,615	1.34301198	$ 33,659.00	$ 3.37
4	Massachusetts	6,451,279	4,429,798	3,919,139	1.294152214	$ 42,123.00	$ 4.21
5	Connecticut	3,474,610	2,375,074	2,064,204	1.277633461	$ 46,417.00	$ 4.64
6	Virginia	7,468,914	5,069,885	4,392,319	1.266878157	$ 36,912.00	$ 3.69
7	California	35,558,419	23,202,576	21,575,797	1.259290324	$ 36,830.00	$ 3.68
8	Illinois	12,645,295	7,999,510	7,529,966	1.228083331	$ 35,957.00	$ 3.60
9	Pennsylvania	12,388,368	8,345,018	6,420,037	1.191848273	$ 33,852.00	$ 3.39
10	New York	19,297,933	12,369,803	9,939,759	1.156059667	$ 38,398.00	$ 3.84
11	Michigan	10,089,305	6,062,886	5,430,637	1.139178863	$ 31,650.00	$ 3.17
12	Ohio	11,464,593	6,677,236	6,188,081	1.122178258	$ 31,617.00	$ 3.16
13	Tennessee	5,916,762	3,294,083	3,171,487	1.092754787	$ 30,297.00	$ 3.03
14	Indiana	6,214,454	3,596,991	2,844,568	1.036544643	$ 30,645.00	$ 3.06
15	Utah	2,438,915	1,228,687	1,229,029	1.007708756	$ 26,827.00	$ 2.68
	Total	159,975,432					
	Average		6,911,284	6,233,604	1.32071	$ 36,603.20	$ 3.66
	Correlation Coefficient -	0.75383809					

Websites used:

http://www.census.gov/popest/states/NST-ann-est.html
http://www.fcc.gov/Bureaus/Common_Carrier/Reports/FCC-State_Link/IAD/trend801.pdf
http://www.bea.gov/newsreleases/relsarchivespi.htm
http://hraunfoss.fcc.gov/edocs_public/attachmatch/DOC-284932A1.pdf
http://www.fcc.gov/Bureaus/Common_Carrier/Reports/FCC-State_Link/IAD/trend605.pdf

Year - 2004		
State	Tele - Accessibility (Fixed + Mobile lines /Population)	Income per Capita /10000
District of Columbia	2.909578541	$ 5.15
New Jersey	1.485752125	$ 4.24
Florida	1.34301198	$ 3.37
Massachusetts	1.294152214	$ 4.21
Connecticut	1.277633461	$ 4.64
Virginia	1.266878157	$ 3.69
California	1.259290324	$ 3.68
Illinois	1.228083331	$ 3.60
Pennsylvania	1.191848273	$ 3.39
New York	1.156059667	$ 3.84
Michigan	1.139178863	$ 3.17
Ohio	1.122178258	$ 3.16
Tennessee	1.092754787	$ 3.03
Indiana	1.036544643	$ 3.06
Utah	1.007708756	$ 2.68
Average	1.32071	
Correlation Coefficient -	0.753838095	

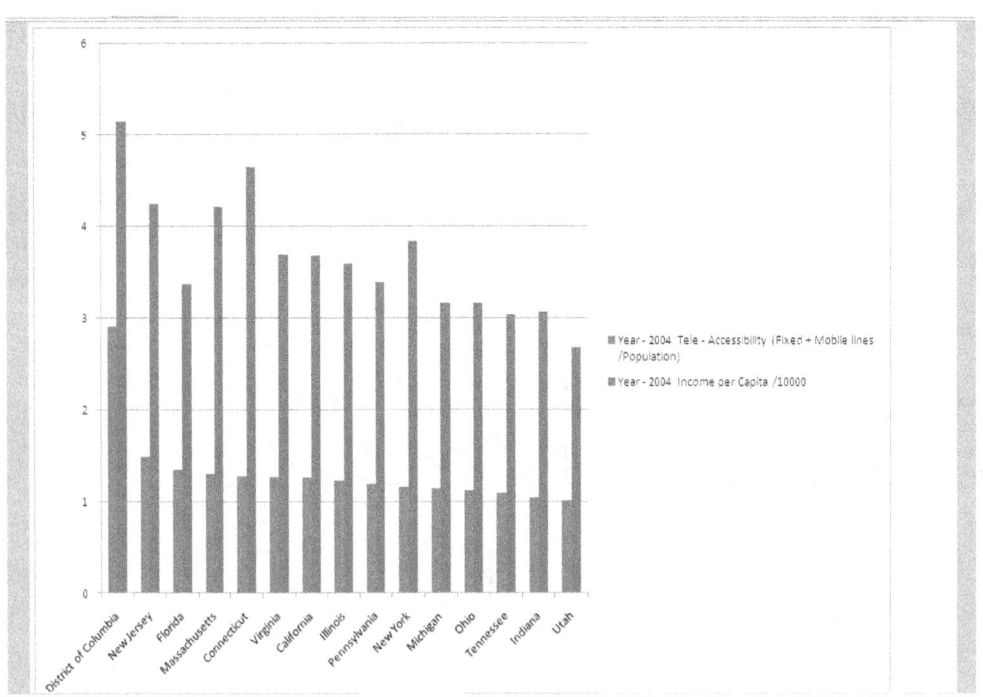

Year - 2004 Tele - Accessibility (Fixed + Mobile lines /Population)

Year - 2004 Income per Capita /10000

	State	Population	Number of Fixed Lines	Number of Mobile Lines	Tele - Accessibility (Fixed + Mobile lines /Population)	Income per Capita	Income per Capita / 10,000
					Year - 2005		
1	District of Columbia	582,049	791,292	752,548	2.652422734	$ 55,268.00	$ 5.53
2	New Jersey	8,621,837	5,983,082	6,233,984	1.416991066	$ 43,994.00	$ 4.40
3	Florida	17,783,868	10,356,866	12,619,929	1.292002111	$ 35,769.00	$ 3.58
4	Connecticut	3,477,416	2,135,021	2,328,966	1.283708075	$ 48,485.00	$ 4.85
5	California	35,795,255	21,285,036	24,572,034	1.281093542	$ 38,670.00	$ 3.87
6	Massachusetts	6,453,031	3,779,199	4,487,601	1.281072414	$ 43,897.00	$ 4.39
7	New York	19,330,891	11,284,418	12,995,534	1.256018256	$ 40,678.00	$ 4.07
8	Illinois	12,674,452	7,323,440	8,227,185	1.226926813	$ 37,168.00	$ 3.72
9	Virginia	7,563,887	4,290,319	4,851,206	1.208575036	$ 38,980.00	$ 3.90
10	Tennessee	5,995,748	3,085,676	4,065,964	1.192785287	$ 31,360.00	$ 3.14
11	Pennsylvania	12,418,161	7,345,150	7,397,397	1.187176346	$ 34,978.00	$ 3.50
12	Michigan	10,090,554	5,688,091	6,229,949	1.181108589	$ 32,265.00	$ 3.23
13	Ohio	11,475,262	6,372,077	6,993,803	1.164755977	$ 32,498.00	$ 3.25
14	Indiana	6,253,120	3,496,667	3,442,612	1.109730662	$ 31,302.00	$ 3.13
15	Utah	2,499,637	1,056,543	1,413,756	0.988263096	$ 28,599.00	$ 2.86
	Total	161,015,168					
	Average		6,284,858	7,107,498	1.31484	$ 38,260.73	$ 3.83

Correlation Coefficient - 0.77669105

Correlation Coefficient - 0.77669105

Websites used:

http://www.census.gov/popest/states/NST-ann-est.html
http://www.fcc.gov/Bureaus/Common_Carrier/Reports/FCC-State_Link/IAD/trend801.pdf
http://www.bea.gov/newsreleases/relsarchivespi.htm
http://hraunfoss.fcc.gov/edocs_public/attachmatch/DOC-284932A1.pdf
http://hraunfoss.fcc.gov/edocs_public/attachmatch/DOC-270407A1.pdf

Year - 2005		
State	Tele - Accessibility (Fixed + Mobile lines /Population)	Income per Capita /10000
District of Columbia	2.652422734	$ 5.53
New Jersey	1.416991066	$ 4.40
Florida	1.292002111	$ 3.58
Connecticut	1.283708075	$ 4.85
California	1.281093542	$ 3.87
Massachusetts	1.281072414	$ 4.39
New York	1.256018256	$ 4.07
Illinois	1.226926813	$ 3.72
Virginia	1.208575036	$ 3.90
Tennessee	1.192785287	$ 3.14
Pennsylvania	1.187176346	$ 3.50
Michigan	1.181108589	$ 3.23
Ohio	1.164755977	$ 3.25
Indiana	1.109730662	$ 3.13
Utah	0.988263096	$ 2.86
Average	1.31484	
Correlation Coefficient -	0.776691046	

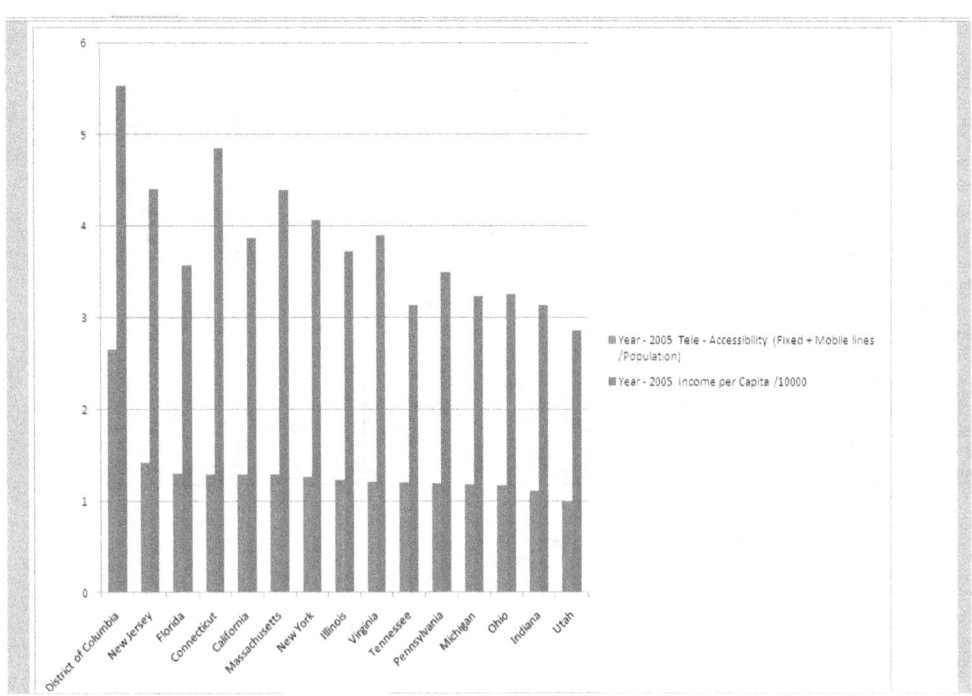

Legend:
■ Year - 2005 Tele - Accessibility (Fixed + Mobile lines /Population)
■ Year - 2005 Income per Capita /10000

		Year - 2006					
	State	Population	Number of Fixed Lines	Number of Mobile Lines	Tele - Accessibility (Fixed + Mobile lines /Population)	Income per Capita	Income per Capita / 10,000
1	District of Columbia	583,978	724,194	878,846	2.745034916	$ 60,080.00	$ 6.01
2	New Jersey	8,623,721	5,048,228	6,953,528	1.391714319	$ 47,655.00	$ 4.77
3	California	35,979,208	19,638,748	27,496,682	1.310074141	$ 41,404.00	$ 4.14
4	Florida	18,088,505	9,214,613	14,176,756	1.293162094	$ 38,308.00	$ 3.83
5	Connecticut	3,485,162	1,854,403	2,582,367	1.273045557	$ 52,702.00	$ 5.27
6	Massachusetts	6,466,399	3,232,105	4,916,500	1.260145716	$ 47,330.00	$ 4.73
7	Tennessee	6,089,453	2,827,951	4,730,704	1.241269947	$ 32,986.00	$ 3.30
8	Illinois	12,718,011	6,463,779	9,147,657	1.227506094	$ 39,549.00	$ 3.95
9	New York	19,356,564	9,183,939	14,573,548	1.227360755	$ 43,973.00	$ 4.40
10	Virginia	7,646,996	3,861,542	5,325,173	1.201349523	$ 41,367.00	$ 4.14
11	Pennsylvania	12,471,142	6,602,383	8,348,713	1.198855406	$ 37,326.00	$ 3.73
12	Ohio	11,492,495	5,433,993	7,939,126	1.163639314	$ 34,093.00	$ 3.41
13	Michigan	10,082,438	4,684,096	6,862,582	1.145226779	$ 33,198.00	$ 3.32
14	Indiana	6,301,700	3,071,465	3,972,560	1.117797578	$ 32,881.00	$ 3.29
15	Utah	2,583,724	977,879	1,649,265	1.016805201	$ 30,320.00	$ 3.03
	Total	161,969,496					
	Average		5,521,288	7,970,267	1.32087	$ 40,878.13	$ 4.09

Correlation Coefficient - 0.74439103

Websites used:

http://www.census.gov/popest/states/NST-ann-est.html
http://www.fcc.gov/Bureaus/Common_Carrier/Reports/FCC-State_Link/IAD/trend801.pdf
http://www.bea.gov/newsreleases/relsarchivespi.htm
http://hraunfoss.fcc.gov/edocs_public/attachmatch/DOC-284932A1.pdf

Year - 2006		
State	Tele - Accessibility (Fixed + Mobile lines /Population)	Income per Capita /10000
District of Columbia	2.745034916	$ 6.01
New Jersey	1.391714319	$ 4.77
California	1.310074141	$ 4.14
Florida	1.293162094	$ 3.83
Connecticut	1.273045557	$ 5.27
Massachusetts	1.260145716	$ 4.73
Tennessee	1.241269947	$ 3.30
Illinois	1.227506094	$ 3.95
New York	1.227360755	$ 4.40
Virginia	1.201349523	$ 4.14
Pennsylvania	1.198855406	$ 3.73
Ohio	1.163639314	$ 3.41
Michigan	1.145226779	$ 3.32
Indiana	1.117797578	$ 3.29
Utah	1.016805201	$ 3.03
Average	1.32087	
Correlation Coefficient -	0.744391033	

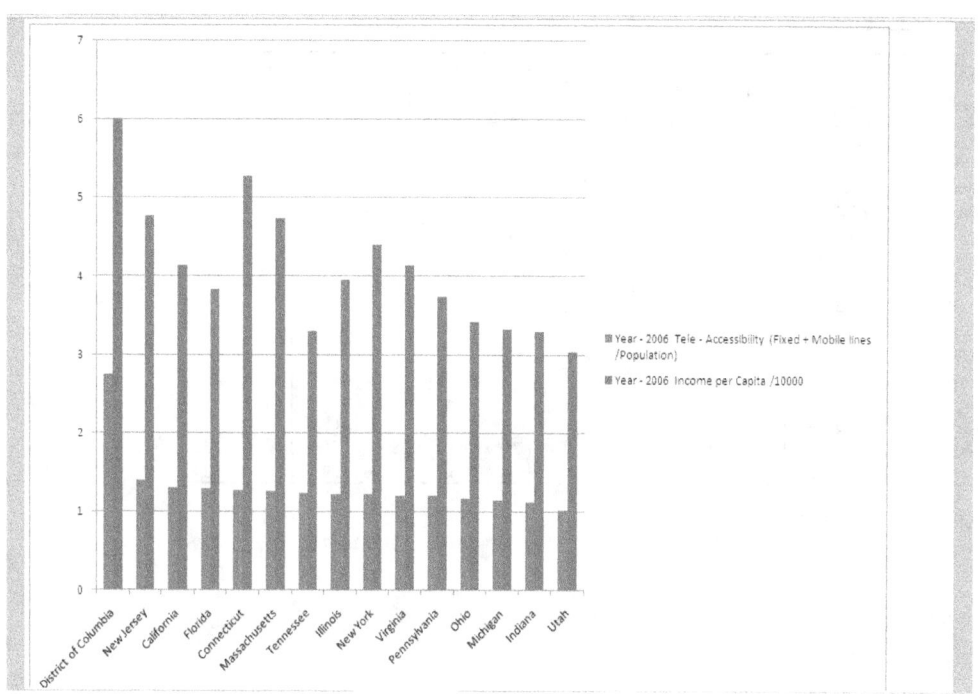

			Year - 2007				
	State	Population	Number of Fixed Lines	Number of Mobile Lines	Tele - Accessibility (Fixed + Mobile lines /Population)	Income per Capita	Income per Capita / 10,000
1	District of Columbia	586,409	969,396	965,816	3.30010624	$ 63,881.00	$ 6.39
2	New Jersey	8,636,043	5,242,137	7,419,289	1.466114284	$ 50,265.00	$ 5.03
3	California	36,226,122	21,383,910	30,203,842	1.424048426	$ 43,221.00	$ 4.32
4	Virginia	7,719,749	4,690,533	6,148,261	1.404034509	$ 43,275.00	$ 4.33
5	Connecticut	3,488,633	2,045,739	2,786,594	1.385165192	$ 55,609.00	$ 5.56
6	Massachusetts	6,499,275	3,695,288	5,289,432	1.382418808	$ 49,885.00	$ 4.99
7	Florida	18,277,888	10,003,949	15,255,433	1.381963934	$ 39,204.00	$ 3.92
8	New York	19,422,777	10,270,594	15,901,378	1.347488673	$ 47,612.00	$ 4.76
9	Pennsylvania	12,522,531	7,473,799	9,200,793	1.331567237	$ 39,058.00	$ 3.91
10	Illinois	12,779,417	6,925,387	9,949,126	1.320444665	$ 41,569.00	$ 4.16
11	Tennessee	6,172,862	3,101,391	4,970,756	1.307683049	$ 34,287.00	$ 3.43
12	Ohio	11,520,815	6,041,991	8,722,523	1.281551175	$ 35,307.00	$ 3.53
13	Michigan	10,050,847	5,041,315	7,333,242	1.231195441	$ 34,188.00	$ 3.42
14	Indiana	6,346,113	3,167,264	4,448,186	1.200018027	$ 33,756.00	$ 3.38
15	Utah	2,663,796	1,106,095	1,874,345	1.118869463	$ 31,739.00	$ 3.17
	Total	162,913,277					
	Average		6,077,253	8,697,934	1.45884	$ 42,857.07	$ 4.29

Correlation Coefficient - 0.7299811

Websites used:

http://www.census.gov/popest/states/NST-ann-est.html
http://www.fcc.gov/Bureaus/Common_Carrier/Reports/FCC-State_Link/IAD/trend801.pdf
http://www.bea.gov/newsreleases/relsarchivespi.htm
http://hraunfoss.fcc.gov/edocs_public/attachmatch/DOC-284932A1.pdf

Year - 2007		
State	Tele - Accessibility (Fixed + Mobile lines /Population)	Income per Capita /10000
District of Columbia	3.30010624	$ 6.39
New Jersey	1.466114284	$ 5.03
California	1.424048426	$ 4.32
Virginia	1.404034509	$ 4.33
Connecticut	1.385165192	$ 5.56
Massachusetts	1.382418808	$ 4.99
Florida	1.381963934	$ 3.92
New York	1.347488673	$ 4.76
Pennsylvania	1.331567237	$ 3.91
Illinois	1.320444665	$ 4.16
Tennessee	1.307683049	$ 3.43
Ohio	1.281551175	$ 3.53
Michigan	1.231195441	$ 3.42
Indiana	1.200018027	$ 3.38
Utah	1.118869463	$ 3.17
Average	1.45884	
Correlation Coefficient -	0.729981104	

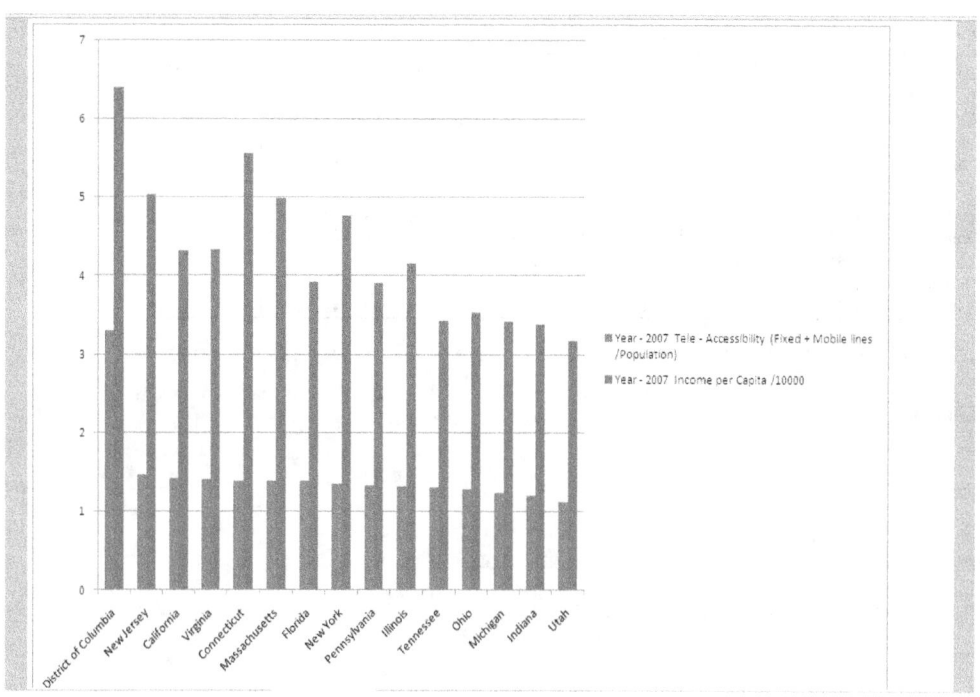

Year - 2007 Tele - Accessibility (Fixed + Mobile lines /Population)

Year - 2007 Income per Capita /10000

Year - 2000 - 2007							
Years	State	Population	Number of Fixed Lines	Number of Mobile Lines	Tele - Accessibility (Fixed + Mobile lines /Population)	Income per Capita	Income per Capita / 10,000
2000	California	33,994,571	24,754,207	12,283,369	1.08951444	$ 32,275.00	$ 3.23
2001	California	34,485,623	23,385,691	14,184,625	1.089448667	$ 32,702.00	$ 3.27
2002	California	34,876,194	24,174,586	16,007,376	1.152131508	$ 33,749.00	$ 3.37
2003	California	35,251,107	23,692,322	18,892,619	1.208045495	$ 34,922.00	$ 3.49
2004	California	35,558,419	23,202,576	21,575,797	1.259290324	$ 36,830.00	$ 3.68
2005	California	35,795,255	21,285,036	24,572,034	1.281093542	$ 38,670.00	$ 3.87
2006	California	35,979,208	19,638,748	27,496,682	1.310074141	$ 41,404.00	$ 4.14
2007	California	36,226,122	21,383,910	30,203,842	1.424048426	$ 43,221.00	$ 4.32
	Average	35,270,812	22,689,635	20,652,043	1.22671	$ 36,721.63	3.67

Correlation Coefficient -   0.97167224

Websites used:

http://www.census.gov/popest/states/NST-ann-est.html
http://www.fcc.gov/Bureaus/Common_Carrier/Reports/FCC-State_Link/IAD/trend801.pdf
http://www.bea.gov/newsreleases/relsarchivespi.htm
http://hraunfoss.fcc.gov/edocs_public/attachmatch/DOC-284932A1.pdf

		California: Year - 2000 - 2007		
Years	State	Tele - Accessibility (Fixed + Mobile lines /Population)	Income per Capita /10000	
2000	California	1.08951444	$	3.23
2001	California	1.089448667	$	3.27
2002	California	1.152131508	$	3.37
2003	California	1.208045495	$	3.49
2004	California	1.259290324	$	3.68
2005	California	1.281093542	$	3.87
2006	California	1.310074141	$	4.14
2007	California	1.424048426	$	4.32

Average	1.22671	
Correlation Coefficient -	0.971672237	

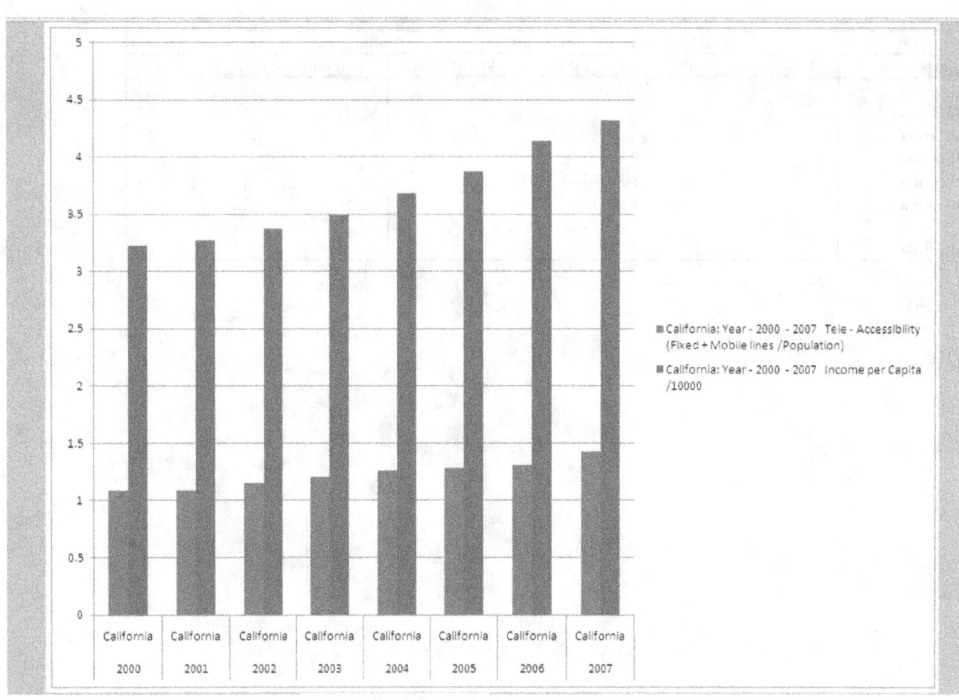

					Tele - Accessibility		Income per	
			Number of	Number of	(Fixed + Mobile lines	Income per		Capita /
Years	State	Population	Fixed Lines	Mobile Lines	/Population)	Capita		10,000
2000	Connecticut	3,411,726	2,574,205	1,136,618	1.087667357	$ 40,640.00	$	4.06
2001	Connecticut	3,428,433	2,406,704	1,418,367	1.115690754	$ 42,435.00	$	4.24
2002	Connecticut	3,448,382	2,499,908	1,577,873	1.182520092	$ 43,173.00	$	4.32
2003	Connecticut	3,467,673	2,449,918	1,791,944	1.223258941	$ 43,730.00	$	4.37
2004	Connecticut	3,474,610	2,375,074	2,064,204	1.277633461	$ 46,417.00	$	4.64
2005	Connecticut	3,477,416	2,135,021	2,328,966	1.283708075	$ 48,485.00	$	4.85
2006	Connecticut	3,485,162	1,854,403	2,582,367	1.273045557	$ 52,702.00	$	5.27
2007	Connecticut	3,488,633	2,045,739	2,786,594	1.385165192	$ 55,609.00	$	5.56
	Average	3,460,254	2,292,622	1,960,867	1.22859	$ 46,648.88		4.66

Year - 2000 - 2007

Correlation Coefficient - 0.90444641

Websites used:

http://www.census.gov/hhes/www/income/income98.html
http://www.fcc.gov/Bureaus/Common_Carrier/Reports/FCC-State_Link/IAD/trend801.pdf
http://hraunfoss.fcc.gov/edocs_public/attachmatch/DOC-284932A1.pdf

Connecticut: Year - 2000 - 2007			
Years	State	Tele - Accessibility (Fixed + Mobile lines /Population)	Income per Capita /10000
2000	Connecticut	1.087667357	$ 4.06
2001	Connecticut	1.115690754	$ 4.24
2002	Connecticut	1.182520092	$ 4.32
2003	Connecticut	1.223258941	$ 4.37
2004	Connecticut	1.277633461	$ 4.64
2005	Connecticut	1.283708075	$ 4.85
2006	Connecticut	1.273045557	$ 5.27
2007	Connecticut	1.385165192	$ 5.56

	Average	1.22859
	Correlation Coefficient -	0.904446406

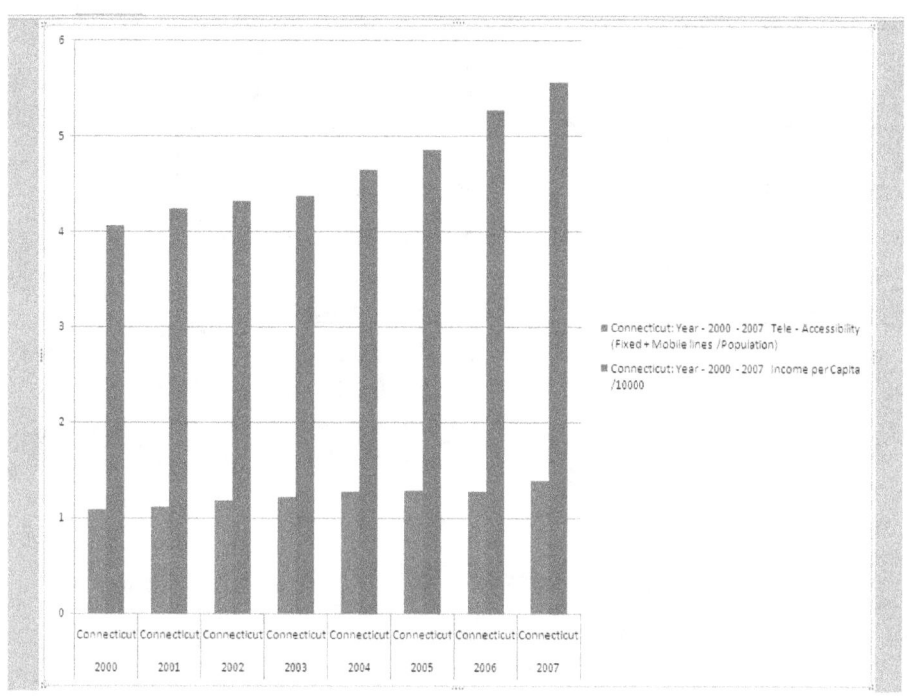

Connecticut: Year - 2000 - 2007 Tele - Accessibility (Fixed + Mobile lines /Population)

Connecticut: Year - 2000 - 2007 Income per Capita /10000

					Tele - Accessibility		Income per
			Number of	Number of	(Fixed + Mobile lines	Income per	Capita /
Years	State	Population	Fixed Lines	Mobile Lines	/Population)	Capita	10,000
2000	District of Columbia	571,744	987,412	333,815	2.310871649	$ 37,383.00	$ 3.74
2001	District of Columbia	578,042	919,587	382,457	2.252507603	$ 40,150.00	$ 4.02
2002	District of Columbia	579,585	992,094	415,399	2.428449667	$ 48,342.00	$ 4.83
2003	District of Columbia	577,777	947,171	520,182	2.539652842	$ 47,529.00	$ 4.75
2004	District of Columbia	579,796	1,131,004	555,958	2.909578541	$ 51,458.00	$ 5.15
2005	District of Columbia	582,049	791,292	752,548	2.652422734	$ 55,268.00	$ 5.53
2006	District of Columbia	583,978	724,194	878,846	2.745034916	$ 60,080.00	$ 6.01
2007	District of Columbia	586,409	969,396	965,816	3.30010624	$ 63,881.00	$ 6.39
	Average	579,923	932,769	600,628	2.64233	$ 50,511.38	5.05

Correlation Coefficient -   0.87000502

Websites used:

http://www.census.gov/popest/states/NST-ann-est.html
http://www.fcc.gov/Bureaus/Common_Carrier/Reports/FCC-State_Link/IAD/trend801.pdf
http://www.bea.gov/newsreleases/relsarchivespi.htm
http://hraunfoss.fcc.gov/edocs_public/attachmatch/DOC-284932A1.pdf

District of Columbia: Year - 2000 - 2007			
Years	State	Tele - Accessibility (Fixed + Mobile lines /Population)	Income per Capita /10000
2000	District of Columbia	2.310871649	$ 3.74
2001	District of Columbia	2.252507603	$ 4.02
2002	District of Columbia	2.428449667	$ 4.83
2003	District of Columbia	2.539652842	$ 4.75
2004	District of Columbia	2.909578541	$ 5.15
2005	District of Columbia	2.652422734	$ 5.53
2006	District of Columbia	2.745034916	$ 6.01
2007	District of Columbia	3.30010624	$ 6.39

Average   2.64233

Correlation Coefficient -   0.87000502

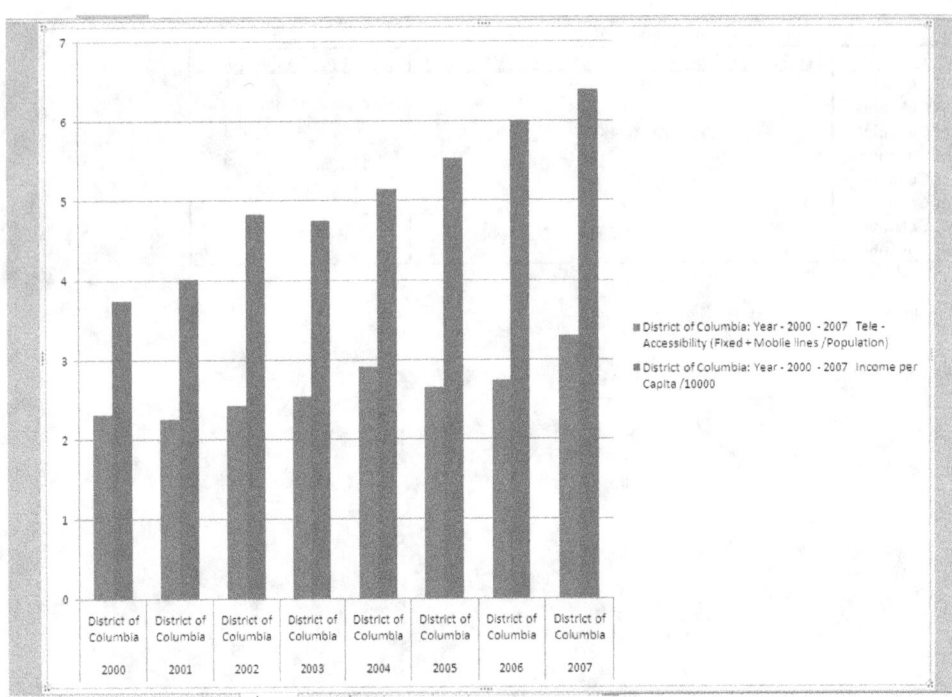

District of Columbia: Year - 2000 - 2007 Tele - Accessibility (Fixed + Mobile lines /Population)

District of Columbia: Year - 2000 - 2007 Income per Capita /10000

					Tele - Accessibility		Income per
Years	State	Population	Number of Fixed Lines	Number of Mobile Lines	(Fixed + Mobile lines /Population)	Income per Capita		Capita / 10,000
2000	Florida	16,047,118	12,104,421	4,983,478	1.064857814	$	28,145.00	$ 2.81
2001	Florida	16,353,869	11,317,933	7,536,670	1.152913907	$	28,947.00	$ 2.89
2002	Florida	16,680,309	11,901,261	8,607,715	1.229532139	$	30,446.00	$ 3.04
2003	Florida	16,981,183	11,671,497	10,252,348	1.291067	$	31,364.00	$ 3.14
2004	Florida	17,375,259	11,418,566	11,916,615	1.34301198	$	33,659.00	$ 3.37
2005	Florida	17,783,868	10,356,866	12,619,929	1.292002111	$	35,769.00	$ 3.58
2006	Florida	18,088,505	9,214,613	14,176,756	1.293162094	$	38,308.00	$ 3.83
2007	Florida	18,277,888	10,003,949	15,255,433	1.381963934	$	39,204.00	$ 3.92
	Average	17,198,500	10,998,638	10,668,618	1.25606	$	33,230.25	3.32

Correlation Coefficient - 0.817447242

Websites used:

http://www.census.gov/popest/states/NST-ann-est.html
http://www.fcc.gov/Bureaus/Common_Carrier/Reports/FCC-State_Link/IAD/trend801.pdf
http://www.bea.gov/newsreleases/relsarchivespi.htm
http://hraunfoss.fcc.gov/edocs_public/attachmatch/DOC-284932A1.pdf

Florida: Year - 2000 - 2007			
Years	State	Tele - Accessibility (Fixed + Mobile lines /Population)	Income per Capita /10000
2000	Florida	1.064857814	$ 2.81
2001	Florida	1.152913907	$ 2.89
2002	Florida	1.229532139	$ 3.04
2003	Florida	1.291067	$ 3.14
2004	Florida	1.34301198	$ 3.37
2005	Florida	1.292002111	$ 3.58
2006	Florida	1.293162094	$ 3.83
2007	Florida	1.381963934	$ 3.92
	Average	1.25606	
	Correlation Coefficient -	0.817447242	

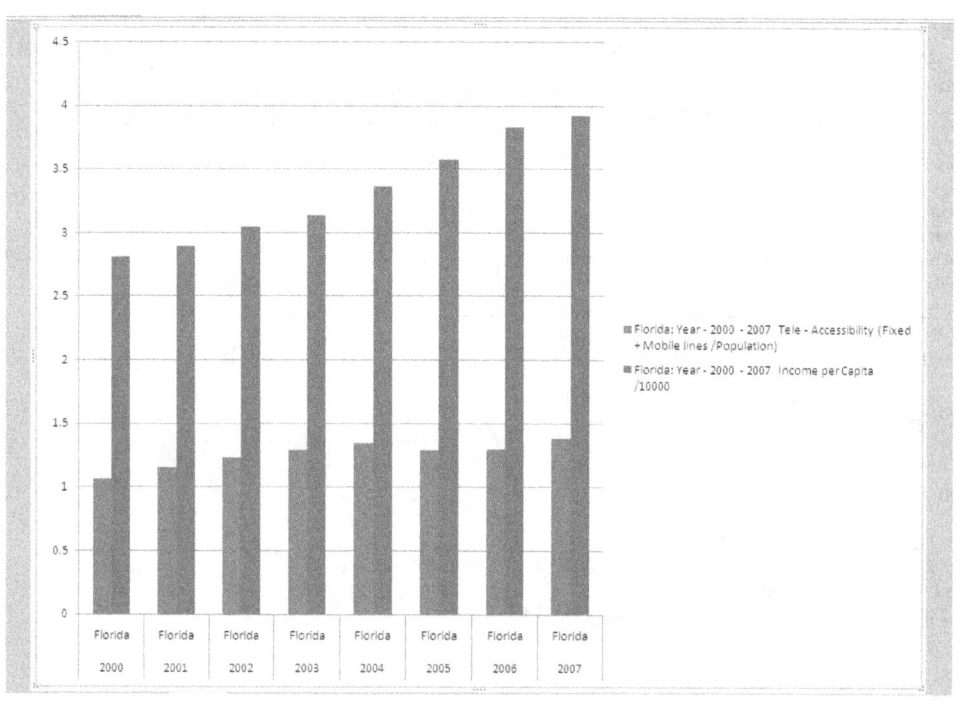

Florida: Year - 2000 - 2007 Tele - Accessibility (Fixed + Mobile lines /Population)

Florida: Year - 2000 - 2007 Income per Capita /10000

Year - 2000 - 2007							
Years	State	Population	Number of Fixed Lines	Number of Mobile Lines	Tele - Accessibility (Fixed + Mobile lines /Population)	Income per Capita	Income per Capita / 10,000
2000	Illinois	12,437,645	8,740,081	4,309,660	1.049213175	$ 32,259.00	$ 3.23
2001	Illinois	12,507,833	8,012,870	5,621,044	1.090030064	$ 33,023.00	$ 3.30
2002	Illinois	12,558,229	8,596,609	5,409,370	1.115282975	$ 33,690.00	$ 3.37
2003	Illinois	12,597,981	8,357,937	6,834,217	1.205919742	$ 34,569.00	$ 3.46
2004	Illinois	12,645,295	7,999,510	7,529,966	1.228083331	$ 35,957.00	$ 3.60
2005	Illinois	12,674,452	7,323,440	8,227,185	1.226926813	$ 37,168.00	$ 3.72
2006	Illinois	12,718,011	6,463,779	9,147,657	1.227506094	$ 39,549.00	$ 3.95
2007	Illinois	12,779,417	6,925,387	9,949,126	1.320444665	$ 41,569.00	$ 4.16
	Average	12,614,858	7,802,452	7,128,528	1.18293	$ 35,973.00	3.60

Correlation Coefficient - 0.909010324

Websites used:

http://www.census.gov/popest/states/NST-ann-est.html
http://www.fcc.gov/Bureaus/Common_Carrier/Reports/FCC-State_Link/IAD/trend801.pdf
http://www.bea.gov/newsreleases/relsarchivespi.htm
http://hraunfoss.fcc.gov/edocs_public/attachmatch/DOC-284932A1.pdf

Illinois: Year - 2000 - 2007			
Years	State	Tele - Accessibility (Fixed + Mobile lines /Population)	Income per Capita /10000
2000	Illinois	1.049213175	$ 3.23
2001	Illinois	1.090030064	$ 3.30
2002	Illinois	1.115282975	$ 3.37
2003	Illinois	1.205919742	$ 3.46
2004	Illinois	1.228083331	$ 3.60
2005	Illinois	1.226926813	$ 3.72
2006	Illinois	1.227506094	$ 3.95
2007	Illinois	1.320444665	$ 4.16

Average	1.18293
Correlation Coefficient -	0.909010324

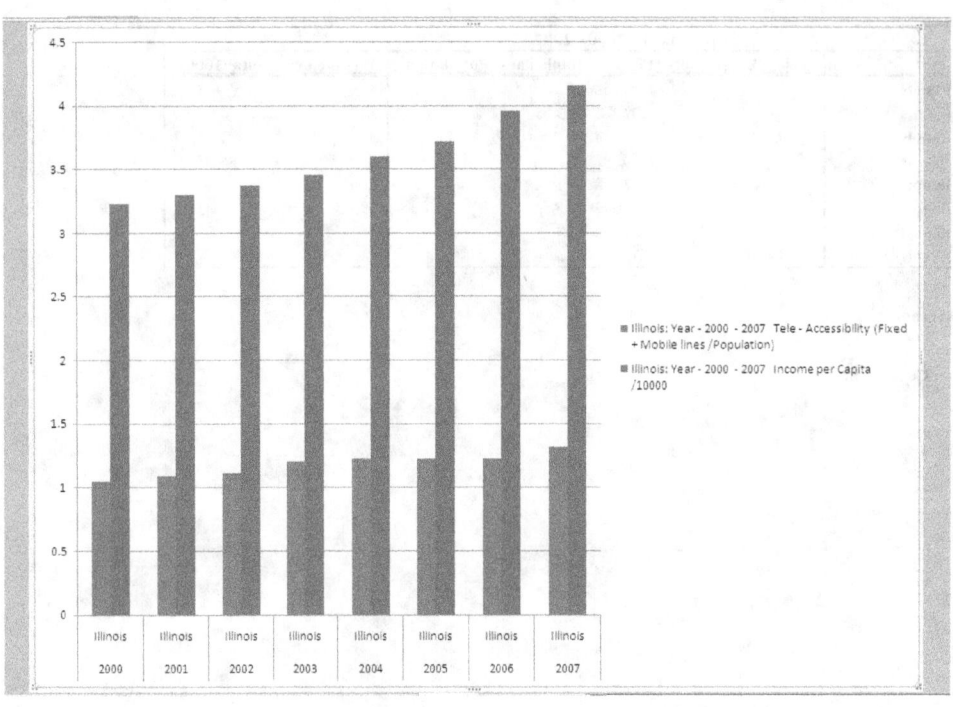

Illinois: Year - 2000 - 2007 Tele - Accessibility (Fixed + Mobile lines /Population)

Illinois: Year - 2000 - 2007 Income per Capita /10000

					Tele - Accessibility		Income per
Years	State	Population	Number of Fixed Lines	Number of Mobile Lines	(Fixed + Mobile lines /Population)	Income per Capita	Capita / 10,000
2000	Indiana	6,091,649	3,753,645	1,717,378	0.898118555	$ 27,011.00	$ 2.70
2001	Indiana	6,124,967	3,803,634	1,781,247	0.911822219	$ 27,783.00	$ 2.78
2002	Indiana	6,149,007	3,744,405	2,032,290	0.939451687	$ 28,783.00	$ 2.88
2003	Indiana	6,181,789	3,675,394	2,456,509	0.991930168	$ 29,588.00	$ 2.96
2004	Indiana	6,214,454	3,596,991	2,844,568	1.036544643	$ 30,645.00	$ 3.06
2005	Indiana	6,253,120	3,496,667	3,442,612	1.109730662	$ 31,302.00	$ 3.13
2006	Indiana	6,301,700	3,071,465	3,972,560	1.117797578	$ 32,881.00	$ 3.29
2007	Indiana	6,346,113	3,167,264	4,448,186	1.200018027	$ 33,756.00	$ 3.38
	Average	6,207,850	3,538,683	2,836,919	1.02568	$ 30,218.63	3.02

Year - 2000 - 2007

Correlation Coefficient - 0.981302157

Websites used:

http://www.census.gov/popest/states/NST-ann-est.html
http://www.fcc.gov/Bureaus/Common_Carrier/Reports/FCC-State_Link/IAD/trend801.pdf
http://www.bea.gov/newsreleases/relsarchivespi.htm
http://hraunfoss.fcc.gov/edocs_public/attachmatch/DOC-284932A1.pdf

Indiana: Year - 2000 - 2007			
Years	State	Tele - Accessibility (Fixed + Mobile lines /Population)	Income per Capita /10000
2000	Indiana	0.898118555	$ 2.70
2001	Indiana	0.911822219	$ 2.78
2002	Indiana	0.939451687	$ 2.88
2003	Indiana	0.991930168	$ 2.96
2004	Indiana	1.036544643	$ 3.06
2005	Indiana	1.109730662	$ 3.13
2006	Indiana	1.117797578	$ 3.29
2007	Indiana	1.200018027	$ 3.38
	Average	1.02568	
	Correlation Coefficient -	0.981302157	

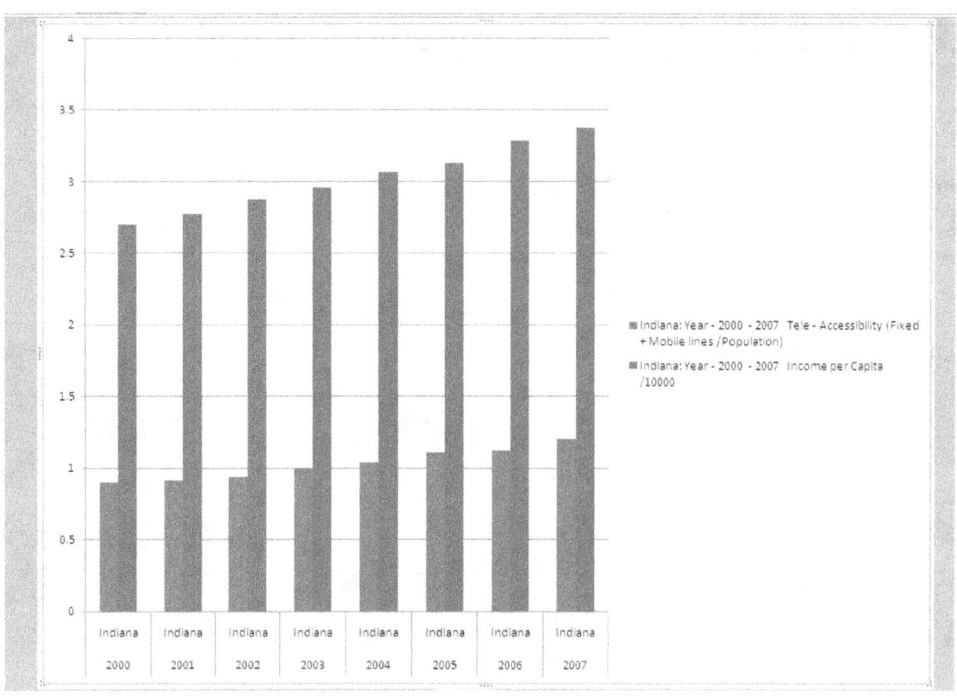

Legend:
- Indiana: Year - 2000 - 2007 Tele - Accessibility (Fixed + Mobile lines /Population)
- Indiana: Year - 2000 - 2007 Income per Capita /10000

Years	State	Population	Number of Fixed Lines	Number of Mobile Lines	Tele - Accessibility (Fixed + Mobile lines /Population)	Income per Capita	Income per Capita / 10,000
					Year - 2000 - 2007		
2000	Massachusetts	6,363,015	4,698,536	2,228,169	1.088588507	$ 37,992.00	$ 3.80
2001	Massachusetts	6,411,730	4,410,394	2,753,685	1.11733947	$ 38,907.00	$ 3.89
2002	Massachusetts	6,440,978	4,501,471	3,289,934	1.209661794	$ 39,815.00	$ 3.98
2003	Massachusetts	6,451,637	4,407,964	3,506,039	1.226665883	$ 40,161.00	$ 4.02
2004	Massachusetts	6,451,279	4,429,798	3,919,139	1.294152214	$ 42,123.00	$ 4.21
2005	Massachusetts	6,453,031	3,779,199	4,487,601	1.281072414	$ 43,897.00	$ 4.39
2006	Massachusetts	6,466,399	3,232,105	4,916,500	1.260145716	$ 47,330.00	$ 4.73
2007	Massachusetts	6,499,275	3,695,288	5,289,432	1.382418808	$ 49,885.00	$ 4.99
	Average	6,442,168	4,144,344	3,798,812	1.23251	$ 42,513.75	4.25

Correlation Coefficient - 0.854092106

Websites used:

http://www.census.gov/popest/states/NST-ann-est.html
http://www.fcc.gov/Bureaus/Common_Carrier/Reports/FCC-State_Link/IAD/trend801.pdf
http://www.bea.gov/newsreleases/relsarchivespi.htm
http://hraunfoss.fcc.gov/edocs_public/attachmatch/DOC-284932A1.pdf

Massachusetts: Year - 2000 - 2007				
Years	State	Tele - Accessibility (Fixed + Mobile lines /Population)	Income per Capita /10000	
2000	Massachusetts	1.088588507	$	3.80
2001	Massachusetts	1.11733947	$	3.89
2002	Massachusetts	1.209661794	$	3.98
2003	Massachusetts	1.226665883	$	4.02
2004	Massachusetts	1.294152214	$	4.21
2005	Massachusetts	1.281072414	$	4.39
2006	Massachusetts	1.260145716	$	4.73
2007	Massachusetts	1.382418808	$	4.99
	Average	1.23251		
	Correlation Coefficient -	0.854092106		

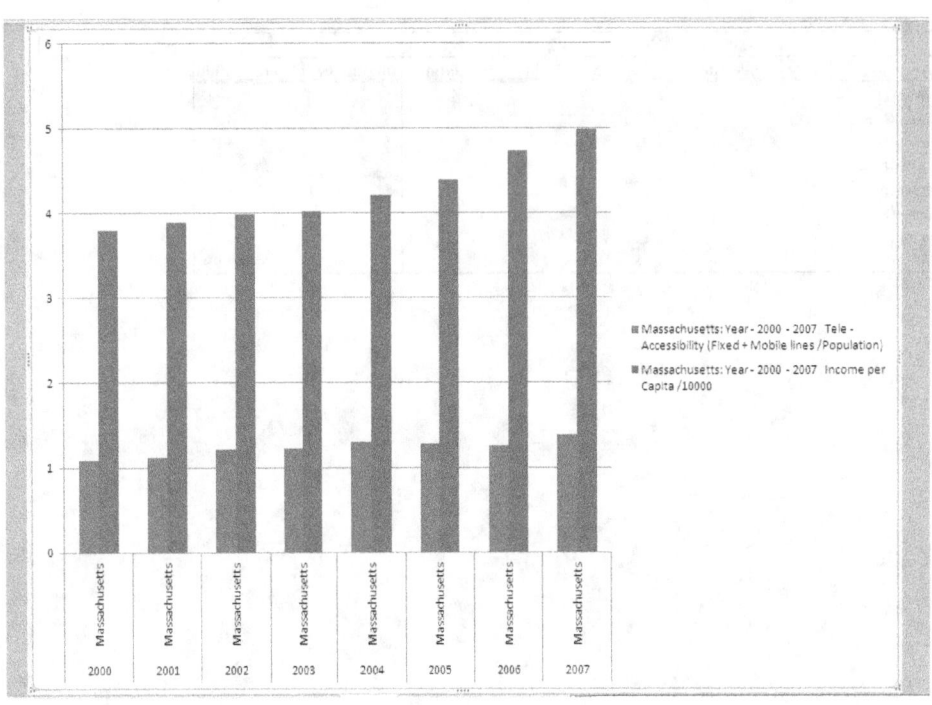

Legend:
- Massachusetts: Year - 2000 - 2007   Tele - Accessibility (Fixed + Mobile lines /Population)
- Massachusetts: Year - 2000 - 2007   Income per Capita /10000

					Tele - Accessibility			Income per	
			Number of	Number of	(Fixed + Mobile lines	Income per		Capita /	
Years	State	Population	Fixed Lines	Mobile Lines	/Population)	Capita		10,000	
2000	Michigan	9,955,308	6,722,255	3,423,535	1.019133712	$	29,612.00	$	2.96
2001	Michigan	10,006,093	6,149,365	4,071,091	1.021423247	$	29,788.00	$	2.98
2002	Michigan	10,038,767	6,536,688	4,758,538	1.12516069	$	30,439.00	$	3.04
2003	Michigan	10,066,351	6,204,267	4,889,269	1.102041445	$	31,214.00	$	3.12
2004	Michigan	10,089,305	6,062,886	5,430,637	1.139178863	$	31,650.00	$	3.17
2005	Michigan	10,090,554	5,688,091	6,229,949	1.181108589	$	32,265.00	$	3.23
2006	Michigan	10,082,438	4,684,096	6,862,582	1.145226779	$	33,198.00	$	3.32
2007	Michigan	10,050,847	5,041,315	7,333,242	1.231195441	$	34,188.00	$	3.42
	Average	10,047,458	5,886,120	5,374,855	1.12056	$	31,544.25		3.15

Year -2000 - 2007

Correlation Coefficient - 0.9047444

Websites used:

http://www.census.gov/popest/states/NST-ann-est.html
http://www.fcc.gov/Bureaus/Common_Carrier/Reports/FCC-State_Link/IAD/trend801.pdf
http://www.bea.gov/newsreleases/relsarchivespi.htm
http://hraunfoss.fcc.gov/edocs_public/attachmatch/DOC-284932A1.pdf

Michigan: Year - 2000 - 2007			
Years	State	Tele - Accessibility (Fixed + Mobile lines /Population)	Income per Capita /10000
2000	Michigan	1.019133712	$ 3.23
2001	Michigan	1.021423247	$ 3.27
2002	Michigan	1.12516069	$ 3.37
2003	Michigan	1.102041445	$ 3.49
2004	Michigan	1.139178863	$ 3.68
2005	Michigan	1.181108589	$ 3.87
2006	Michigan	1.145226779	$ 4.14
2007	Michigan	1.231195441	$ 4.32
	Average	1.12056	
	Correlation Coefficient -	0.904744405	

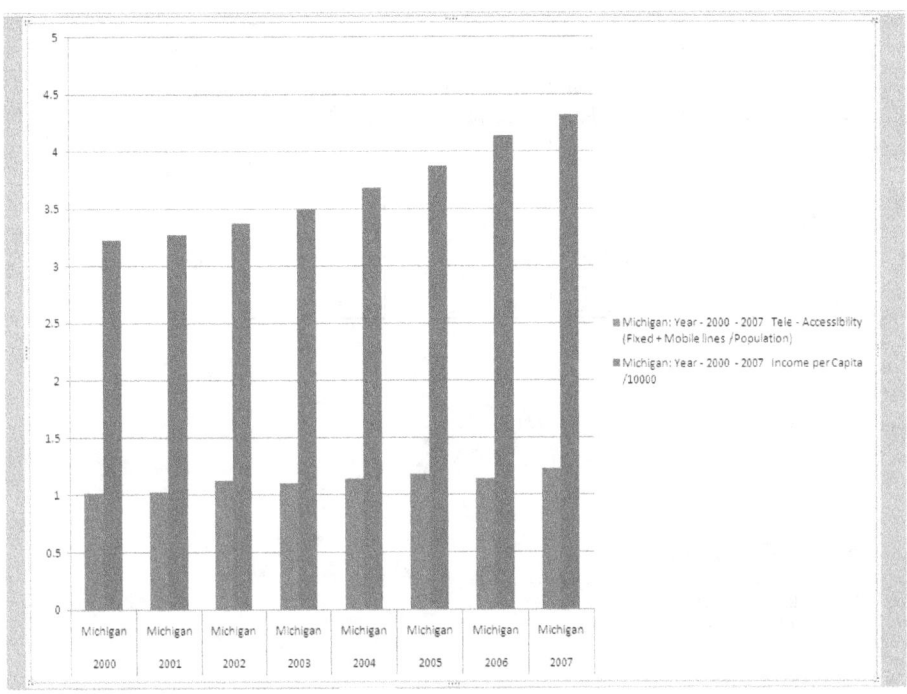

Legend:
- Michigan: Year - 2000 - 2007 Tele - Accessibility (Fixed + Mobile lines /Population)
- Michigan: Year - 2000 - 2007 Income per Capita /10000

			Number of Fixed Lines	Number of Mobile Lines	Tele - Accessibility (Fixed + Mobile lines /Population)	Income per Capita		Income per Capita / 10,000	
					Year -2000 - 2007				
Years	State	Population	Number of Fixed Lines	Number of Mobile Lines	Tele - Accessibility (Fixed + Mobile lines /Population)	Income per Capita		Income per Capita / 10,000	
2000	New Jersey	8,430,921	7,000,131	2,750,024	1.156475669	$	36,983.00	$	3.70
2001	New Jersey	8,489,469	6,923,410	3,896,778	1.274542377	$	38,509.00	$	3.85
2002	New Jersey	8,544,115	6,580,282	4,531,457	1.300513745	$	40,427.00	$	4.04
2003	New Jersey	8,583,481	6,399,743	5,392,240	1.373799627	$	40,504.00	$	4.05
2004	New Jersey	8,611,530	6,468,140	6,326,459	1.485752125	$	42,406.00	$	4.24
2005	New Jersey	8,621,837	5,983,082	6,233,984	1.416991066	$	43,994.00	$	4.40
2006	New Jersey	8,623,721	5,048,228	6,953,528	1.391714319	$	47,655.00	$	4.77
2007	New Jersey	8,636,043	5,242,137	7,419,289	1.466114284	$	50,265.00	$	5.03
	Average	8,567,640	6,205,644	5,437,970	1.35824	$	42,592.88		4.26

Correlation Coefficient - 0.755744069

Websites used:

http://www.census.gov/popest/states/NST-ann-est.html

http://www.fcc.gov/Bureaus/Common_Carrier/Reports/FCC-State_Link/IAD/trend801.pdf

http://www.bea.gov/newsreleases/relsarchivespi.htm

http://hraunfoss.fcc.gov/edocs_public/attachmatch/DOC-284932A1.pdf

New Jersey: Year - 2000 - 2007			
Years	State	Tele - Accessibility (Fixed + Mobile lines /Population)	Income per Capita /10000
2000	New Jersey	1.156475669	$ 3.70
2001	New Jersey	1.274542377	$ 3.85
2002	New Jersey	1.300513745	$ 4.04
2003	New Jersey	1.373799627	$ 4.05
2004	New Jersey	1.485752125	$ 4.24
2005	New Jersey	1.416991066	$ 4.40
2006	New Jersey	1.391714319	$ 4.77
2007	New Jersey	1.466114284	$ 5.03
	Average	1.35824	
	Correlation Coefficient -	0.755744069	

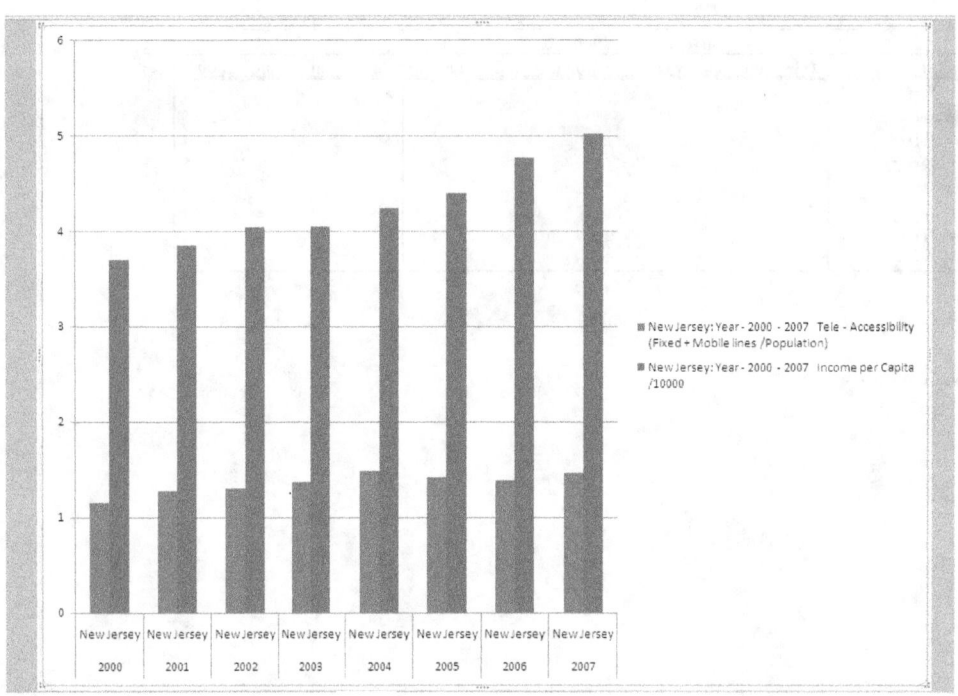

Legend:
- New Jersey: Year - 2000 - 2007 Tele - Accessibility (Fixed + Mobile lines /Population)
- New Jersey: Year - 2000 - 2007 Income per Capita /10000

					Year -2000 - 2007		
Years	State	Population	Number of Fixed Lines	Number of Mobile Lines	Tele - Accessibility (Fixed + Mobile lines /Population)	Income per Capita	Income per Capita / 10,000
2000	New York	18,998,044	13,689,883	5,016,524	0.984649104	$ 34,547.00	$ 3.45
2001	New York	19,088,978	13,076,558	6,749,096	1.038591694	$ 36,019.00	$ 3.60
2002	New York	19,161,873	12,821,422	7,915,526	1.082198384	$ 36,574.00	$ 3.66
2003	New York	19,231,101	12,498,312	8,829,070	1.109004731	$ 36,165.00	$ 3.62
2004	New York	19,297,933	12,369,803	9,939,759	1.156059667	$ 38,398.00	$ 3.84
2005	New York	19,330,891	11,284,418	12,995,534	1.256018256	$ 40,678.00	$ 4.07
2006	New York	19,356,564	9,183,939	14,573,548	1.227360755	$ 43,973.00	$ 4.40
2007	New York	19,422,777	10,270,594	15,901,378	1.347488673	$ 47,612.00	$ 4.76
	Average	19,236,020	11,899,366	10,240,054	1.15017	$ 39,245.75	3.92

Correlation Coefficient - 0.935778535

Websites used:

http://www.census.gov/popest/states/NST-ann-est.html
http://www.fcc.gov/Bureaus/Common_Carrier/Reports/FCC-State_Link/IAD/trend801.pdf
http://www.bea.gov/newsreleases/relsarchivespi.htm
http://hraunfoss.fcc.gov/edocs_public/attachmatch/DOC-284932A1.pdf

New York: Year - 2000 - 2007			
Years	State	Tele - Accessibility (Fixed + Mobile lines /Population)	Income per Capita /10000
2000	New York	1.08951444	$ 3.45
2001	New York	1.089448667	$ 3.60
2002	New York	1.152131508	$ 3.66
2003	New York	1.208045495	$ 3.62
2004	New York	1.259290324	$ 3.84
2005	New York	1.281093542	$ 4.07
2006	New York	1.310074141	$ 4.40
2007	New York	1.424048426	$ 4.76
	Average	1.15017	
	Correlation Coefficient -	0.939458907	

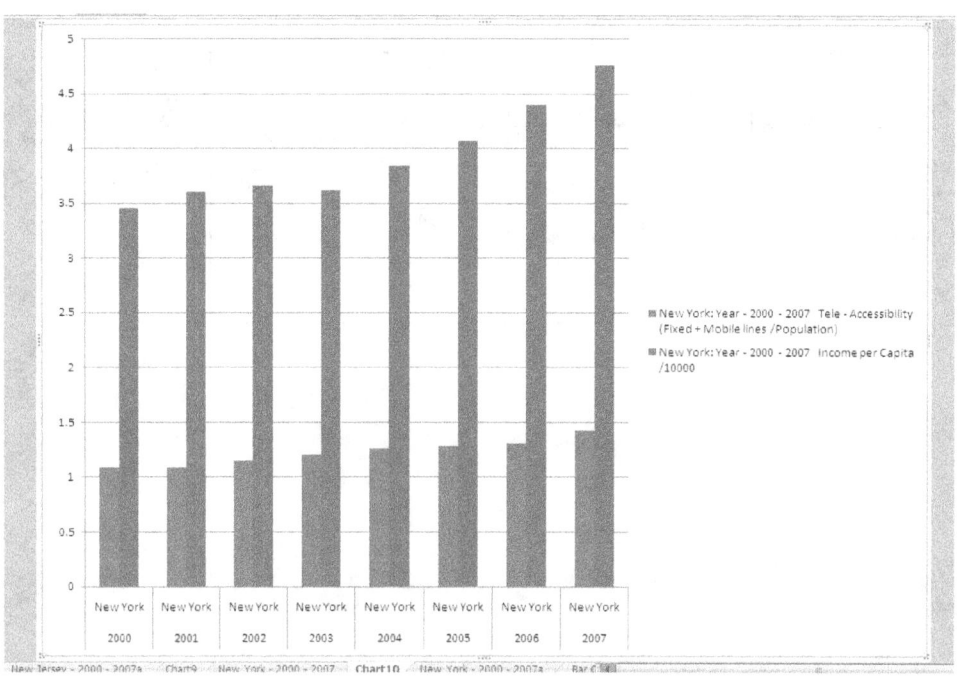

Legend:
■ New York: Year - 2000 - 2007   Tele - Accessibility (Fixed + Mobile lines /Population)
■ New York: Year - 2000 - 2007   Income per Capita /10000

| | | | | | Tele - Accessibility | | Income per | | Income per | |
Years	State	Population	Number of Fixed Lines	Number of Mobile Lines	(Fixed + Mobile lines /Population)		Income per Capita		Capita / 10,000	
2000	Pennsylvania	12,285,504	8,871,784	3,850,372	1.03554205	$	29,539.00	$	2.95	
2001	Pennsylvania	12,299,533	8,301,408	4,378,216	1.030902881	$	30,720.00	$	3.07	
2002	Pennsylvania	12,326,302	8,573,098	4,987,067	1.100100014	$	31,998.00	$	3.20	
2003	Pennsylvania	12,357,524	8,261,544	5,681,653	1.128316401	$	32,427.00	$	3.24	
2004	Pennsylvania	12,388,368	8,345,018	6,420,037	1.191848273	$	33,852.00	$	3.39	
2005	Pennsylvania	12,418,161	7,345,150	7,397,397	1.187176346	$	34,978.00	$	3.50	
2006	Pennsylvania	12,471,142	6,602,383	8,348,713	1.198855406	$	37,326.00	$	3.73	
2007	Pennsylvania	12,522,531	7,473,799	9,200,793	1.331567237	$	39,058.00	$	3.91	
	Average	12,383,633	7,971,773	6,283,031	1.15054	$	33,737.25		3.37	

Year -2000 - 2007

Correlation Coefficient - 0.950827235

Websites used:

http://www.census.gov/popest/states/NST-ann-est.html
http://www.fcc.gov/Bureaus/Common_Carrier/Reports/FCC-State_Link/IAD/trend801.pdf
http://www.bea.gov/newsreleases/relsarchivespi.htm
http://hraunfoss.fcc.gov/edocs_public/attachmatch/DOC-284932A1.pdf

Pennsylvania: Year - 2000 - 2007			
Years	State	Tele - Accessibility (Fixed + Mobile lines /Population)	Income per Capita /10000
2000	Pennsylvania	1.03554205	$ 2.95
2001	Pennsylvania	1.030902881	$ 3.07
2002	Pennsylvania	1.100100014	$ 3.20
2003	Pennsylvania	1.128316401	$ 3.24
2004	Pennsylvania	1.191848273	$ 3.39
2005	Pennsylvania	1.187176346	$ 3.50
2006	Pennsylvania	1.198855406	$ 3.73
2007	Pennsylvania	1.331567237	$ 3.91
	Average	1.15054	
	Correlation Coefficient -	0.950827235	

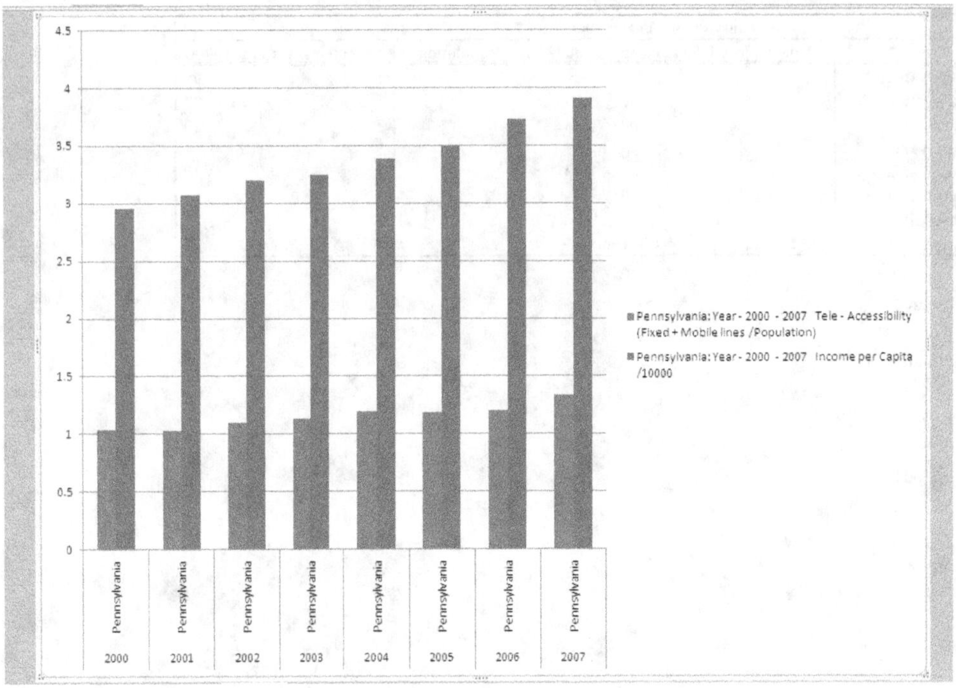

4.5
4
3.5
3
2.5
2
1.5
1
0.5
0
Pennsylvania
2000
2001
2002
2003
2004
2005
2006
2007
Pennsylvania: Year - 2000 - 2007 Tele - Accessibility (Fixed + Mobile lines /Population)
Pennsylvania: Year - 2000 - 2007 Income per Capita /10000

					Tele - Accessibility				Income per
Years	State	Population	Number of Fixed Lines	Number of Mobile Lines	(Fixed + Mobile lines /Population)		Income per Capita		Capita / 10,000	
2000	Tennessee	5,703,243	3,525,455	1,876,444	0.947162693		$	26,239.00	$	2.62
2001	Tennessee	5,755,443	3,385,953	2,251,208	0.979448671		$	26,988.00	$	2.70
2002	Tennessee	5,803,306	3,474,219	2,660,068	1.057033181		$	28,455.00	$	2.85
2003	Tennessee	5,856,522	3,388,799	2,800,735	1.056861735		$	29,026.00	$	2.90
2004	Tennessee	5,916,762	3,294,083	3,171,487	1.092754787		$	30,297.00	$	3.03
2005	Tennessee	5,995,748	3,085,676	4,065,964	1.192785287		$	31,360.00	$	3.14
2006	Tennessee	6,089,453	2,827,951	4,730,704	1.241269947		$	32,986.00	$	3.30
2007	Tennessee	6,172,862	3,101,391	4,970,756	1.307683049		$	34,287.00	$	3.43
	Average	5,911,667	3,260,441	3,315,921	1.10937		$	29,954.75		3.00

Title above table: **Year -2000 - 2007**

Correlation Coefficient - 0.99181703

Websites used:

http://www.census.gov/popest/states/NST-ann-est.html
http://www.fcc.gov/Bureaus/Common_Carrier/Reports/FCC-State_Link/IAD/trend801.pdf
http://www.bea.gov/newsreleases/relsarchivespi.htm
http://hraunfoss.fcc.gov/edocs_public/attachmatch/DOC-284932A1.pdf

Tennessee: Year - 2000 - 2007			
Years	State	Tele - Accessibility (Fixed + Mobile lines /Population)	Income per Capita /10000
2000	Tennessee	0.947162693	$ 2.62
2001	Tennessee	0.979448671	$ 2.70
2002	Tennessee	1.057033181	$ 2.85
2003	Tennessee	1.056861735	$ 2.90
2004	Tennessee	1.092754787	$ 3.03
2005	Tennessee	1.192785287	$ 3.14
2006	Tennessee	1.241269947	$ 3.30
2007	Tennessee	1.307683049	$ 3.43
	Average	1.10937	
	Correlation Coefficient -	0.991817027	

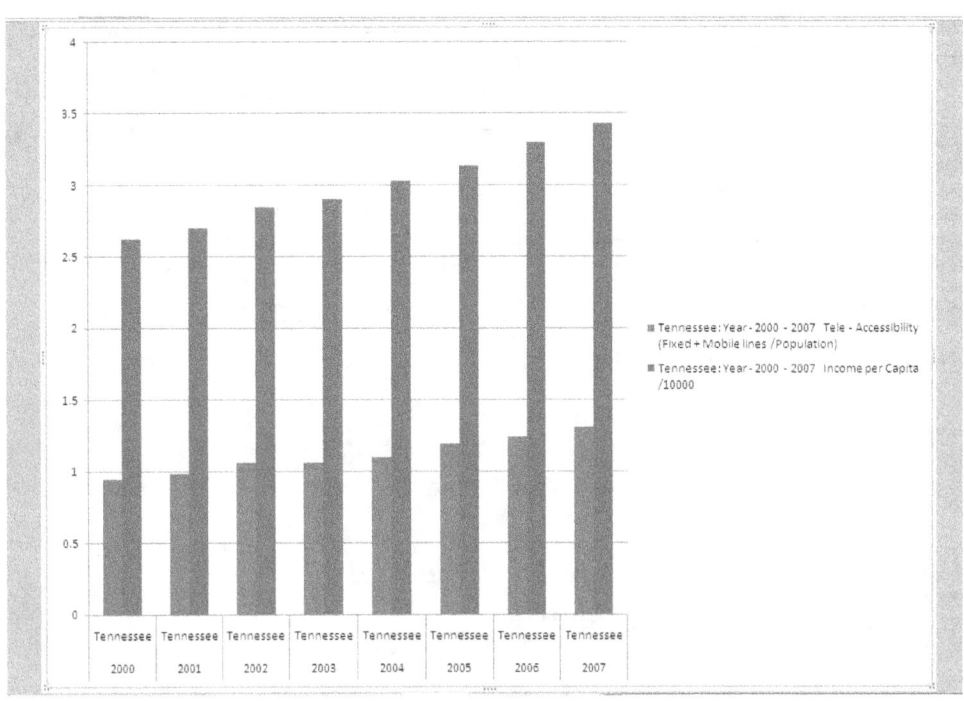

Legend:
■ Tennessee: Year - 2000 - 2007   Tele - Accessibility (Fixed + Mobile lines /Population)
■ Tennessee: Year - 2000 - 2007   Income per Capita /10000

Year - 2000 - 2007							
Years	State	Population	Number of Fixed Lines	Number of Mobile Lines	Tele - Accessibility (Fixed + Mobile lines /Population)	Income per Capita	Income per Capita / 10,000
2000	Virginia	7,104,533	4,469,865	2,447,687	0.973681451	$ 31,162.00	$ 3.12
2001	Virginia	7,191,304	4,760,302	3,059,420	1.08738582	$ 32,431.00	$ 3.24
2002	Virginia	7,283,541	4,902,153	3,429,450	1.14389457	$ 33,671.00	$ 3.37
2003	Virginia	7,373,694	4,759,521	3,879,582	1.171611271	$ 35,029.00	$ 3.50
2004	Virginia	7,468,914	5,069,885	4,392,319	1.266878157	$ 36,912.00	$ 3.69
2005	Virginia	7,563,887	4,290,319	4,851,206	1.208575036	$ 38,980.00	$ 3.90
2006	Virginia	7,646,996	3,861,542	5,325,173	1.201349523	$ 41,367.00	$ 4.14
2007	Virginia	7,719,749	4,690,533	6,148,261	1.404034509	$ 43,275.00	$ 4.33
	Average	7,419,077	4,600,515	4,191,637	1.18218	$ 36,603.38	3.66

Correlation Coefficient - 0.86839009

Websites used:

http://www.census.gov/popest/states/NST-ann-est.html
http://www.fcc.gov/Bureaus/Common_Carrier/Reports/FCC-State_Link/IAD/trend801.pdf
http://www.bea.gov/newsreleases/relsarchivespi.htm
http://hraunfoss.fcc.gov/edocs_public/attachmatch/DOC-284932A1.pdf

Virginia: Year - 2000 - 2007				
Years	State	Tele - Accessibility (Fixed + Mobile lines /Population)	Income per Capita /10000	
2000	Virginia	0.973681451	$	3.12
2001	Virginia	1.08738582	$	3.24
2002	Virginia	1.14389457	$	3.37
2003	Virginia	1.171611271	$	3.50
2004	Virginia	1.266878157	$	3.69
2005	Virginia	1.208575036	$	3.90
2006	Virginia	1.201349523	$	4.14
2007	Virginia	1.404034509	$	4.33
	Average	1.18218		
	Correlation Coefficient -	0.868390086		

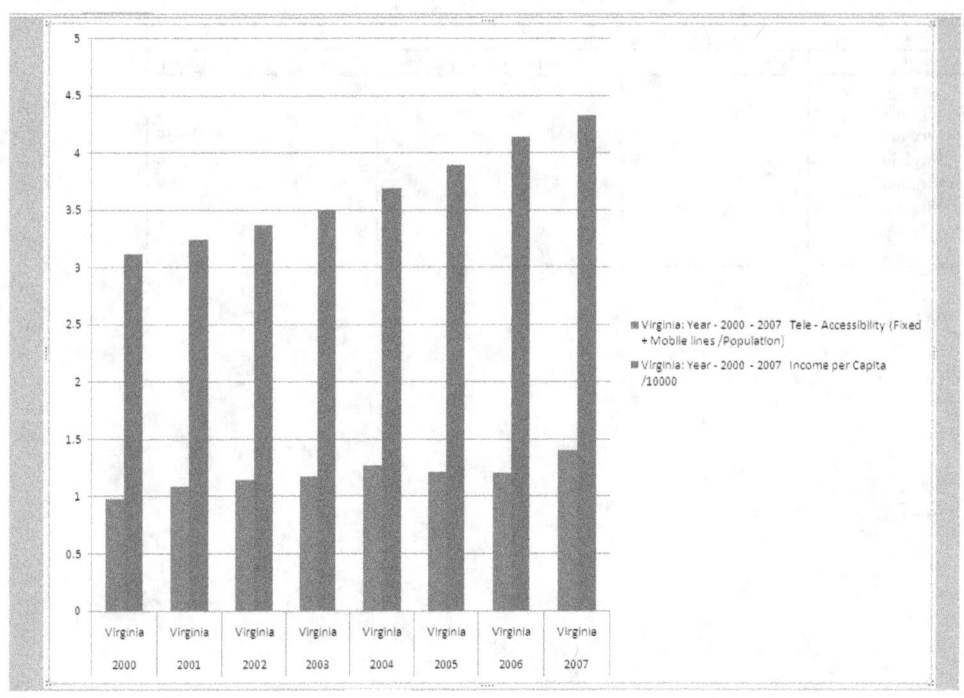

Legend:
- Virginia: Year - 2000 - 2007 Tele - Accessibility (Fixed + Mobile lines /Population)
- Virginia: Year - 2000 - 2007 Income per Capita /10000

					Tele - Accessibility		Income per
					(Fixed + Mobile lines	Income per	Capita /
Years	State	Population	Number of Fixed Lines	Number of Mobile Lines	/Population)	Capita	10,000
2000	Utah	2,244,314	1,286,615	692,006	0.88161505	$23,907.00	$ 2.39
2001	Utah	2,291,250	1,172,443	833,492	0.875476268	$24,180.00	$ 2.42
2002	Utah	2,334,473	1,269,413	970,854	0.95964571	$24,977.00	$ 2.50
2003	Utah	2,379,938	1,254,259	1,094,563	0.98692571	$25,830.00	$ 2.58
2004	Utah	2,438,915	1,228,687	1,229,029	1.007708756	$26,827.00	$ 2.68
2005	Utah	2,499,637	1,056,543	1,413,756	0.988263096	$28,599.00	$ 2.86
2006	Utah	2,583,724	977,879	1,649,265	1.016805201	$30,320.00	$ 3.03
2007	Utah	2,663,796	1,106,095	1,874,345	1.118869463	$31,739.00	$ 3.17
	Average	2,429,506	1,168,992	1,219,664	0.97941	$27,047.38	2.70

Year - 2000 - 2007

Correlation Coefficient - 0.896774465

Websites used:

http://www.census.gov/popest/states/NST-ann-est.html
http://www.fcc.gov/Bureaus/Common_Carrier/Reports/FCC-State_Link/IAD/trend801.pdf
http://www.bea.gov/newsreleases/relsarchivespi.htm
http://hraunfoss.fcc.gov/edocs_public/attachmatch/DOC-284932A1.pdf

Utah: Year - 2000 - 2007			
Years	State	Tele - Accessibility (Fixed + Mobile lines /Population)	Income per Capita /10000
2000	Utah	0.88161505	S 2.39
2001	Utah	0.875476268	S 2.42
2002	Utah	0.95964571	S 2.50
2003	Utah	0.98692571	S 2.58
2004	Utah	1.007708756	S 2.68
2005	Utah	0.988263096	S 2.86
2006	Utah	1.016805201	S 3.03
2007	Utah	1.118869463	S 3.17

Average   0.97941

Correlation Coefficient -   0.896774465

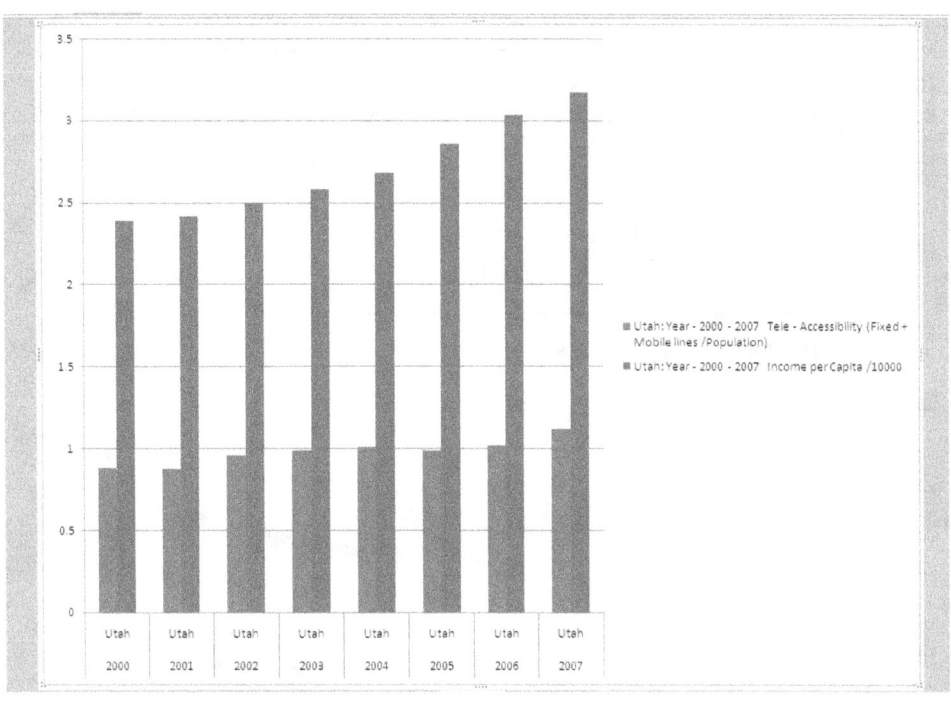

Utah:Year - 2000 - 2007 Tele - Accessibility (Fixed + Mobile lines /Population)

Utah:Year - 2000 - 2007 Income per Capita /10000

					Tele - Accessibility		Income per
			Number of	Number of	(Fixed + Mobile lines	Income per	Capita /
Years	State	Population	Fixed Lines	Mobile Lines	/Population)	Capita	10,000
2000	Ohio	11,363,844	7,211,041	3,278,960	0.923103221	$ 28,400.00	$ 2.84
2001	Ohio	11,396,874	7,053,650	4,255,934	0.992340882	$ 28,816.00	$ 2.88
2002	Ohio	11,420,981	7,057,674	4,887,376	1.045886514	$ 29,944.00	$ 2.99
2003	Ohio	11,445,180	6,885,788	5,659,459	1.096116182	$ 30,698.00	$ 3.07
2004	Ohio	11,464,593	6,677,236	6,188,081	1.122178258	$ 31,617.00	$ 3.16
2005	Ohio	11,475,262	6,372,077	6,993,803	1.164755977	$ 32,498.00	$ 3.25
2006	Ohio	11,492,495	5,433,993	7,939,126	1.163639314	$ 34,093.00	$ 3.41
2007	Ohio	11,520,815	6,041,991	8,722,523	1.281551175	$ 35,307.00	$ 3.53
	Average	11,447,506	6,591,681	5,990,658	1.09870	$ 31,421.63	3.14

Year - 2000 - 2007

Correlation Coefficient - 0.962908826

Websites used:

http://www.census.gov/popest/states/NST-ann-est.html
http://www.fcc.gov/Bureaus/Common_Carrier/Reports/FCC-State_Link/IAD/trend801.pdf
http://www.bea.gov/newsreleases/relsarchivespi.htm
http://hraunfoss.fcc.gov/edocs_public/attachmatch/DOC-284932A1.pdf

Ohio: Year - 2000 - 2007			
Years	State	Tele - Accessibility (Fixed + Mobile lines /Population)	Income per Capita /10000
2000	Ohio	0.923103221	$ 2.84
2001	Ohio	0.992340882	$ 2.88
2002	Ohio	1.045886514	$ 2.99
2003	Ohio	1.096116182	$ 3.07
2004	Ohio	1.122178258	$ 3.16
2005	Ohio	1.164755977	$ 3.25
2006	Ohio	1.163639314	$ 3.41
2007	Ohio	1.281551175	$ 3.53

	Average	1.09870
	Correlation Coefficient -	0.962908826

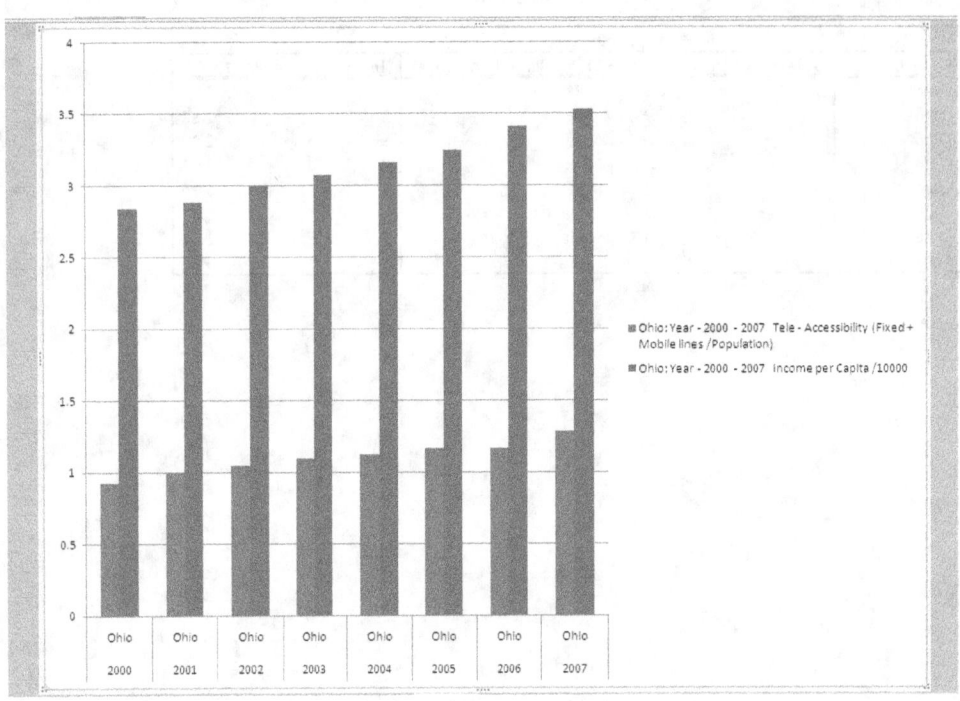

Legend:
- Ohio: Year - 2000 - 2007 Tele - Accessibility (Fixed + Mobile lines /Population)
- Ohio: Year - 2000 - 2007 Income per Capita /10000

Contact Information

If you would like to contact us to ask questions, provide comments, or obtain information about possibly consulting or speaking engagements, please write a letter to the following postal or e-mailing address.

P.O. Box N-9122,

Nassau, Bahamas

New Providence

Or

E-mail address(s)

paulacumberbatch@yahoo.com or pauldelancy42@hotmail.com

Take Care,

God Bless,

Paul Anthony Cumberbatch Jr.